DISCRETE MATHEMATICS AND ITS APPLICATIONS
Series Editor KENNETH H. ROSEN

A JAVA LIBRARY OF GRAPH ALGORITHMS AND OPTIMIZATION

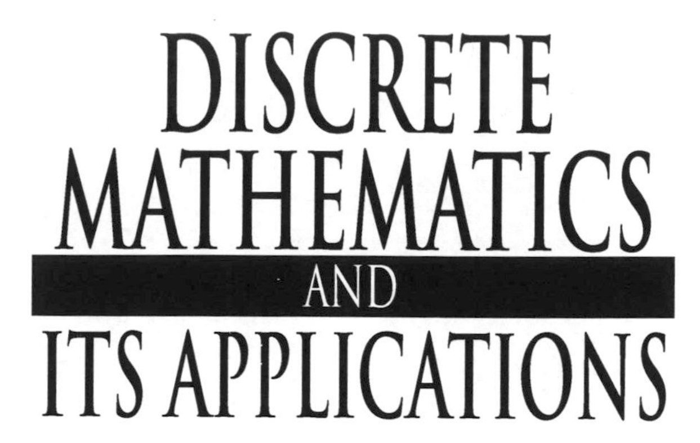

Series Editor

Kenneth H. Rosen, Ph.D.

Juergen Bierbrauer, Introduction to Coding Theory

Kun-Mao Chao and Bang Ye Wu, Spanning Trees and Optimization Problems

Charalambos A. Charalambides, Enumerative Combinatorics

Henri Cohen, Gerhard Frey, et al., Handbook of Elliptic and Hyperelliptic Curve Cryptography

Charles J. Colbourn and Jeffrey H. Dinitz, Handbook of Combinatorial Designs, Second Edition

Steven Furino, Ying Miao, and Jianxing Yin, Frames and Resolvable Designs: Uses, Constructions, and Existence

Randy Goldberg and Lance Riek, A Practical Handbook of Speech Coders

Jacob E. Goodman and Joseph O'Rourke, Handbook of Discrete and Computational Geometry, Second Edition

Jonathan L. Gross and Jay Yellen, Graph Theory and Its Applications, Second Edition

Jonathan L. Gross and Jay Yellen, Handbook of Graph Theory

Darrel R. Hankerson, Greg A. Harris, and Peter D. Johnson, Introduction to Information Theory and Data Compression, Second Edition

Daryl D. Harms, Miroslav Kraetzl, Charles J. Colbourn, and John S. Devitt, Network Reliability: Experiments with a Symbolic Algebra Environment

Leslie Hogben, Handbook of Linear Algebra

Derek F. Holt with Bettina Eick and Eamonn A. O'Brien, Handbook of Computational Group Theory

David M. Jackson and Terry I. Visentin, An Atlas of Smaller Maps in Orientable and Nonorientable Surfaces

Richard E. Klima, Neil P. Sigmon, and Ernest L. Stitzinger, Applications of Abstract Algebra with Maple™ and MATLAB®, Second Edition

Patrick Knupp and Kambiz Salari, Verification of Computer Codes in Computational Science and Engineering

William Kocay and Donald L. Kreher, Graphs, Algorithms, and Optimization

Donald L. Kreher and Douglas R. Stinson, Combinatorial Algorithms: Generation Enumeration and Search

Continued Titles

Charles C. Lindner and Christopher A. Rodgers, Design Theory

Hang T. Lau, A Java Library of Graph Algorithms and Optimization

Alfred J. Menezes, Paul C. van Oorschot, and Scott A. Vanstone, Handbook of Applied Cryptography

Richard A. Mollin, Algebraic Number Theory

Richard A. Mollin, Codes: The Guide to Secrecy from Ancient to Modern Times

Richard A. Mollin, Fundamental Number Theory with Applications

Richard A. Mollin, An Introduction to Cryptography, Second Edition

Richard A. Mollin, Quadratics

Richard A. Mollin, RSA and Public-Key Cryptography

Carlos J. Moreno and Samuel S. Wagstaff, Jr., Sums of Squares of Integers

Dingyi Pei, Authentication Codes and Combinatorial Designs

Kenneth H. Rosen, Handbook of Discrete and Combinatorial Mathematics

Douglas R. Shier and K.T. Wallenius, Applied Mathematical Modeling: A Multidisciplinary Approach

Jörn Steuding, Diophantine Analysis

Douglas R. Stinson, Cryptography: Theory and Practice, Third Edition

Roberto Togneri and Christopher J. deSilva, Fundamentals of Information Theory and Coding Design

Lawrence C. Washington, Elliptic Curves: Number Theory and Cryptography

DISCRETE MATHEMATICS AND ITS APPLICATIONS
Series Editor KENNETH H. ROSEN

A JAVA LIBRARY OF GRAPH ALGORITHMS AND OPTIMIZATION

HANG T. LAU

CRC Press is an imprint of the
Taylor & Francis Group, an informa business
A CHAPMAN & HALL BOOK

Cover design by Vill Mak.

CRC Press
Taylor & Francis Group
6000 Broken Sound Parkway NW, Suite 300
Boca Raton, FL 33487-2742

First issued in paperback 2019

ISBN-13: 978-1-58488-718-8 (hbk)
ISBN-13: 978-0-367-39013-6 (pbk)

Library of Congress Cataloging-in-Publication Data

Lau, H. T. (Hang Tong), 1952-
A Java library of graph algorithms and optimization / Hang T. Lau.
p. cm. -- (Discrete mathematics and its applications)
Includes bibliographical references and index.
ISBN-13: 978-1-58488-718-8 (acid-free paper)
ISBN-10: 1-58488-718-4 (acid-free paper)
1. Java (Computer program language) 2. Computer algorithms. 3. Combinatorial optimization. I. Title. II. Series.

QA76.73.J38L362 2007
005.13'3--dc22 2006024036

Visit the Taylor & Francis Web site at
http://www.taylorandfrancis.com

and the CRC Press Web site at
http://www.crcpress.com

To my wife, Helen,
and our children Matthew, Lawrence, and Tabia
for being understanding of all the times that
I had monopolized the kitchen table with my computer and notes,
or missed goodnight kisses.

The Author

Hang T. Lau is the associate director of the Department of Career and Management Studies, Centre for Continuing Education, McGill University, Montreal, Canada. Dr. Lau has over twenty years of industrial experience in research and development of telecommunications.

Contents

Introduction

Combinatorial optimization has been widely used in applications of different areas. In particular, algorithms in graph theory and mathematical programming have been developed over many years. They often appear in books and journals. Some of the well-known algorithms have been implemented in various computer programming languages. In some cases, the ready-to-use computer codes are generally not easily accessible.

This book attempts to collect some of the most studied graph algorithms and optimization procedures together and serve as a convenient source of program library. The choice of the topics and solution methods is based on the author's interests.

In recent years, Java has become increasingly popular and is often taught as the first programming language in universities. Java code is highly portable and platform independent. Furthermore, the Java Development Kit is freely available. These are the prime reasons in using Java as the programming language for the library in this book.

The main objective of this book is to provide the source code of a library of Java programs that implement some of the most popular graph algorithms and some mathematical programming procedures. The library of programs is primarily intended to be used for educational and experimental purposes. Through the list of simple parameters, programs can be invoked with minimal effort for problem-solving without much concern for their underlying methodology and implementation.

Every chapter is self-contained and largely independent. Each topic starts with a problem description and an outline of the solution procedure. References given in Appendix A of the book should be consulted for all details of the algorithms. The parameters of each procedure are described in detail, followed by the source code listing, and a small test example illustrating the usage of the code. Standard graph-theoretic terminology can be found in texts such as [BM76, H69]. For convenience, Appendix B gives the definitions of most graph-theoretic terms used in this book.

The programs have been tested by using the Java 2 Platform Standard Edition Development Kit 5.0 version 1.5.0.02 on a personal computer with the Windows XP operating system.

1. Random Graph Generation

1.1 Random Permutation of n Objects

Most of the random graph generation procedures in this chapter make use of a subprogram that will generate a random permutation of n objects. A random permutation of the set of integers {1, 2, ..., n} is produced by a sequence of random interchanges.

Procedure parameters:

void randomPermutation (*n, ran, perm*)

n: int;
entry: number of objects to be permuted.

ran: java.util.Random;
entry: a pseudorandom number generator of type *java.util.Random* that has already been initialized by a "seed" of type *long*.

perm: int[*n*+1];
exit: *perm[i]* is the random permutation at *i*, *i* = 1,2,...,*n*.

```
public static void randomPermutation(int n, Random ran, int perm[])
{
  int i,j,k;

  for (i=1; i<=n; i++)
    perm[i] = i;
  for (i=1; i<=n; i++) {
    j = (int)(i + ran.nextDouble() * (n + 1 - i));
    k = perm[i];
    perm[i] = perm[j];
    perm[j] = k;
  }
}
```

Example:

Generate a random permutation of 9 elements.

```
package GraphAlgorithms;
import java.util.Random;

public class Test_randomPermutation extends Object {

  public static void main(String args[]) {

    int n = 9;
    long seed = 1;
    int perm[] = new int[n+1];
```

```
    Random ran = new Random(seed);

    GraphAlgo.randomPermutation(n, ran, perm);
    System.out.println("A random permuation of " + n + " elements:");
    for (int i=1; i<=n; i++)
      System.out.print(" " + perm[i]);
    System.out.println();
  }
}
```

Output:

```
A random permuation of 9 elements:
 7 5 4 2 9 6 3 1 8
```

1.2 Random Graph

The following procedure generates a random graph with some given number of nodes and edges. The type of random graph can be one of the following:

a. Simple (no loops or parallel edges) or nonsimple
b. Directed or undirected
c. Directed acyclic
d. Weighted or unweighted

In generating a random graph, pairs of random integers in the appropriate range are generated. For directed graphs, the pairs are ordered. For a simple graph, loops and parallel edges are excluded. In the case of a directed acyclic graph, a random permutation of n objects (perm[1], perm[2], ..., perm[n]) is first generated. Then m edges in the form of (perm[i],perm[j]) with i < j are generated.

Procedure parameters:

int randomGraph (*n, m, seed, simple, directed, acyclic, weighted, minweight, maxweight, nodei, nodej, weight*)

randomGraph: int;
exit: the method returns the following error code:
0: solution found with normal execution
1: value of *m* is too large, should be at most $n*(n-1)/2$ for simple undirected graph or directed acyclic graph, and $n*(n-1)$ for simple directed graph.

n: int;
entry: number of nodes of the graph.
Nodes of the graph are labeled from 1 to *n*.

m: int;
entry: number of edges of the graph.
If *m* is greater than the maximum number of edges in a graph then a complete graph is generated.

seed: long;
entry: seed for initializing the random number generator.

simple: boolean;
entry: *simple* = true if the graph is simple, false otherwise.

directed: boolean;
entry: *directed* = true if the graph is directed, false otherwise.

acyclic: boolean;
entry: *acyclic* = true if it is a directed acyclic graph, false otherwise.

weighted: boolean;
entry: *weighted* = true if the graph is weighted, false otherwise.

minweight: int;
entry: minimum weight of the edges;
if *weighted* = false then this value is ignored.

maxweight: int;
entry: maximum weight of the edges;
if *weighted* = false then this value is ignored.

nodei, nodej: int[m+1];
exit: the i-th edge is from node *nodei[i]* to node *nodej[i]*,
for i = 1,2,...,m.

weight: int[m+1];
exit: *weight[i]* is the weight of the i-th edge, for i = 1,2,...,m;
if *weighted* = false then this array is ignored.

```
public static int randomGraph(int n, int m, long seed, boolean simple,
      boolean directed, boolean acyclic, boolean weighted, int minweight,
      int maxweight, int nodei[], int nodej[], int weight[])
{
  int maxedges,nodea,nodeb,numedges,temp;
  int dagpermute[] = new int[n + 1];
  boolean adj[][] = new boolean[n+1][n+1];
  Random ran = new Random(seed);

  // initialize the adjacency matrix
  for (nodea=1; nodea<=n; nodea++)
    for (nodeb=1; nodeb<=n; nodeb++)
      adj[nodea][nodeb] = false;
  numedges = 0;
  // check for valid input data
  if (simple) {
    maxedges = n * (n - 1);
    if (!directed) maxedges /= 2;
    if (m > maxedges) return 1;
  }
  if (acyclic) {
    maxedges = (n * (n - 1)) / 2;
    if (m > maxedges) return 1;
    randomPermutation(n,ran,dagpermute);
  }
  while (numedges < m) {
```

```
    nodea = ran.nextInt(n) + 1;
    nodeb = ran.nextInt(n) + 1;
    if (simple || acyclic)
      if (nodea == nodeb) continue;
    if ((simple && (!directed)) || acyclic)
      if (nodea > nodeb) {
        temp = nodea;
        nodea = nodeb;
        nodeb = temp;
      }
    if (acyclic) {
      nodea = dagpermute[nodea];
      nodeb = dagpermute[nodeb];
    }
    if ((!simple) || (simple && (!adj[nodea][nodeb]))) {
      numedges++;
      nodei[numedges] = nodea;
      nodej[numedges] = nodeb;
      adj[nodea][nodeb] = true;
      if (weighted)
        weight[numedges] = (int)(minweight +
              ran.nextDouble() * (maxweight + 1 - minweight));
    }
  }
  return 0;
}
```

Example:

Generate a random simple undirected nonacyclic weighted graph with 5 nodes and 8 edges.

```
package GraphAlgorithms;

public class Test_randomGraph extends Object {

  public static void main(String args[]) {

    int k;
    int n = 5;
    int m = 8;
    long seed = 1;
    boolean simple=true, directed=false, acyclic=false, weighted=true;
    int minweight = 90;
    int maxweight = 99;
    int nodei[] = new int[m+1];
    int nodej[] = new int[m+1];
    int weight[] = new int[m+1];

    k = GraphAlgo.randomGraph(n,m,seed,simple,directed,acyclic,
```

```
              weighted,minweight,maxweight,nodei,nodej,weight);
    if (k != 0)
      System.out.println("Invalid input data, error code = " + k);
    else {
      System.out.println("List of edges:\n    from to  weight");
      for (k=1; k<=m; k++)
        System.out.println("      " + nodei[k] + "    " + nodej[k] +
                                   "     " + weight[k]);
    }
  }
}
```

Output:

```
List of edges:
   from to  weight
     1   4     94
     2   5     99
     4   5     99
     3   5     99
     2   3     93
     1   5     96
     1   3     91
     1   2     95
```

1.3 Random Bipartite Graph

A *bipartite graph* is an undirected graph where the nodes can be partitioned into two sets such that no edge connects nodes in the same set. The following procedure generates a random simple bipartite graph. It is optional to specify the required number of edges. The method generates random edges connecting the nodes of the two sets.

Procedure parameters:

void randomBipartiteGraph (*n1, n2, m, seed, nodei, nodej*)

n1: int;
entry: number of nodes in the first set of the bipartite graph.

n2: int;
entry: number of nodes in the second set of the bipartite graph.
Nodes of the graph are labeled from 1 to *n1*+*n2*.

m: int;
entry: required number of edges of the graph.
If *m* is zero then a random bipartite graph will be generated with random edges connecting the nodes in the first set to

		the nodes in the second set, and the total number of edges will be returned in *nodei[0]*.
seed:	long;	
	entry:	seed for initializing the random number generator.
nodei, nodej:	int[*m*+1] for $m > 0$; int[*n1***n2*+1] for $m = 0$;	
	exit:	The i-th edge is from node *nodei[i]* to node *nodej[i]*, for i = 1,2,..., *nodei[0]*.

```
public static void randomBipartiteGraph(int n1, int n2, int m,
                          long seed, int nodei[], int nodej[])
{
  int n,nodea,nodeb,nodec,numedges;
  boolean adj[][] = new boolean[n1+n2+1][n1+n2+1];
  boolean temp;
  Random ran = new Random(seed);

  n = n1 + n2;
  // initialize the adjacency matrix
  for (nodea=1; nodea<=n; nodea++)
    for (nodeb=1; nodeb<=n; nodeb++)
      adj[nodea][nodeb] = false;

  if (m != 0) {
    if (m > n1 * n2) m = n1 * n2;
    numedges = 0;
    // generate a simple bipartite graph with exactly m edges
    while (numedges < m) {
      // generate a random integer in interval [1, n1]
      nodea = (int)(1 + ran.nextDouble() * n1);
      // generate a random integer in interval [n1+1, n]
      nodeb = (int)(n1 + 1 + ran.nextDouble() * n2);
      if (!adj[nodea][nodeb]) {
        // add the edge (nodei,nodej)
        adj[nodea][nodeb] = adj[nodeb][nodea] = true;
        numedges++;
      }
    }
  }
  else {
    // generate a random adjacency matrix with edges from
    // nodes of group [1, n1] to nodes of group [n1+1, n]
    for (nodea=1; nodea<=n1; nodea++)
      for (nodeb=n1+1; nodeb<=n; nodeb++)
        adj[nodea][nodeb] = adj[nodeb][nodea] =
                   (ran.nextInt(2) == 0) ? false : true;
  }
  // random permutation of rows and columns
  for (nodea=1; nodea<=n; nodea++) {
    nodec = (int)(nodea + ran.nextDouble() * (n + 1 - nodea));
    for (nodeb=1; nodeb<=n; nodeb++) {
```

```
      temp = adj[nodec][nodeb];
      adj[nodec][nodeb] = adj[nodea][nodeb];
      adj[nodea][nodeb] = temp;
    }
    for (nodeb=1; nodeb<=n; nodeb++) {
      temp = adj[nodeb][nodec];
      adj[nodeb][nodec] = adj[nodeb][nodea];
      adj[nodeb][nodea] = temp;
    }
  }
  numedges = 0;
  for (nodea=1; nodea<=n; nodea++)
    for (nodeb=nodea+1; nodeb<=n; nodeb++)
      if (adj[nodea][nodeb]) {
        numedges++;
        nodei[numedges] = nodea;
        nodej[numedges] = nodeb;
      }
  nodei[0] = numedges;
}
```

Example:

Generate a random bipartite graph of 6 edges with 3 nodes in the first set and 4 nodes in the second set.

```
package GraphAlgorithms;

public class Test_randomBipartiteGraph extends Object {

  public static void main(String args[]) {

    int n1 = 3;
    int n2 = 4;
    int m = 6;
    long seed = 1;
    int nodei[] = new int[m+1];
    int nodej[] = new int[m+1];

    GraphAlgo.randomBipartiteGraph(n1,n2,m,seed,nodei,nodej);
    System.out.println("List of edges:\n   from to");
    for (int k=1; k<=nodei[0]; k++)
      System.out.println("      " + nodei[k] + "    " + nodej[k]);
  }
}
```

Output:

```
List of edges:
   from to
     1   2
     1   6
     3   6
     4   5
     5   6
     5   7
```

1.4 Random Regular Graph

The following procedure generates a random simple undirected regular graph with a given number of nodes *n* and the degree *d* of each node. An implementation in programming language C is given in [JK91].

Step 1. Start with a graph with n nodes and no edges.

Step 2. If every node in the graph is of degree *d* then stop.

Step 3. Randomly choose two nonadjacent nodes u and v, each of degree less than *d*. If such nodes u and v are not found then proceed to Step 5.

Step 4. Add the edge (u,v) to the graph, and go back to Step 2.

Step 5. If there is one node of degree less than *d* then proceed to Step 7.

Step 6. From the graph, randomly choose two adjacent nodes r and s each of degree less than *d*. Randomly choose two adjacent nodes p and q such that p is not adjacent to r, and q is not adjacent to s. Remove the edge (p,q) and add the two edges (p,r) and (q,s) to the graph. Go back to Step 2.

Step 7. There is exactly one node r of degree less than *d*. Randomly choose two adjacent nodes p and q that are not adjacent to r. Remove the edge (p,q) and add the two edges (p,r) and (q,r) to the graph. Go back to Step 2.

Procedure parameters:

int randomRegularGraph (*n, degree, seed, nodei, nodej*)

randomRegularGraph: int;
- exit: the method returns the following error code:
  - 0: solution found with normal execution
  - 1: invalid input, if *degree* is odd then *n* must be even
  - 2: value of *n* should be greater than *degree*

n: int;
- entry: number of nodes of the graph. Nodes of the graph are labeled from 1 to *n*.

degree: int;
- entry: required degree of each node. If the value of *degree* is odd then the value *n* must be even.

seed: long;
- entry: seed for initializing the random number generator.

nodei, nodej: int[(*n**degree*)/2 + 1];
exit: the i-th edge is from node *nodei[i]* to node *nodej[i]*, for i = 1, 2, ..., (*n**degree*)/2.

```
public static int randomRegularGraph(int n, int degree, long seed,
                                     int nodei[], int nodej[])
{
  int i,j,numedges,p,q,r=0,s=0,u,v=0;
  int permute[] = new int[n + 1];
  int deg[] = new int[n + 1];
  boolean adj[][] = new boolean[n+1][n+1];
  boolean more;
  Random ran = new Random(seed);

  // initialize the adjacency matrix
  for (i=1; i<=n; i++)
    for (j=1; j<=n; j++)
      adj[i][j] = false;
  // initialize the degree of each node
  for (i=1; i<=n; i++)
    deg[i] = 0;
  // check input data consistency
  if ((degree % 2) != 0)
    if ((n % 2) != 0) return 1;
  if (n <= degree) return 2;
  // generate the regular graph
  iterate:
  while (true) {
    randomPermutation(n,ran,permute);
    more = false;
    // find two non-adjacent nodes each has less than required degree
    u = 0;
    for (i=1; i<=n; i++)
      if (deg[permute[i]] < degree) {
         v = permute[i];
         more = true;
         for (j=i+1; j<=n; j++) {
           if (deg[permute[j]] < degree) {
             u = permute[j];
             if (!adj[v][u]) {
               // add edge (u,v) to the random graph
               adj[v][u] = adj[u][v] = true;
               deg[v]++;
               deg[u]++;
               continue iterate;
             }
             else {
               // both r & s are less than the required degree
               r = v;
               s = u;
```

```
          }
        }
      }
    }
  if (!more) break;
  if (u == 0) {
    r = v;
    // node r has less than the required degree,
    // find two adjacent nodes p and q non-adjacent to r.
    for (i=1; i<=n-1; i++) {
      p = permute[i];
      if (r != p)
        if (!adj[r][p])
          for (j=i+1; j<=n; j++) {
            q = permute[j];
            if (q != r)
              if (adj[p][q] && (!adj[r][q])) {
                // add edges (r,p) & (r,q), delete edge (p,q)
                adj[r][p] = adj[p][r] = true;
                adj[r][q] = adj[q][r] = true;
                adj[p][q] = adj[q][p] = false;
                deg[r]++;
                deg[r]++;
                continue iterate;
              }
          }
    }
  }
  else {
    // nodes r and s of less than required degree, find two
    // adjacent nodes p & q such that (p,r) & (q,s) are not edges.
    for (i=1; i<=n; i++) {
      p = permute[i];
      if ((p != r) && (p != s))
        if (!adj[r][p])
          for (j=1; j<=n; j++) {
            q = permute[j];
            if ((q != r) && (q != s))
              if (adj[p][q] && (!adj[s][q])) {
                // remove edge (p,q), add edges (p,r) & (q,s)
                adj[p][q] = adj[q][p] = false;
                adj[r][p] = adj[p][r] = true;
                adj[s][q] = adj[q][s] = true;
                deg[r]++;
                deg[s]++;
                continue iterate;
              }
          }
    }
  }
```

```
  }
  numedges = 0;
  for (i=1; i<=n; i++)
    for (j=i+1; j<=n; j++)
      if (adj[i][j]) {
        numedges++;
        nodei[numedges] = i;
        nodej[numedges] = j;
      }
  return 0;
}
```

Example:

Generate a random regular graph with 8 nodes of degree 3.

```
package GraphAlgorithms;

public class Test_randomRegularGraph extends Object {

  public static void main(String args[]) {

    int k;
    int n = 8;
    int degree = 3;
    long seed = 1;
    int edges = (n * degree) / 2;
    int nodei[] = new int[edges+1];
    int nodej[] = new int[edges+1];

    k = GraphAlgo.randomRegularGraph(n,degree,seed,nodei,nodej);
    if (k != 0)
      System.out.println("invalid input data, error code = " + k);
    else {
      System.out.print("List of edges:\n from: ");
      for (k=1; k<=edges; k++)
        System.out.print("  " + nodei[k]);
      System.out.print("\n   to: ");
      for (k=1; k<=edges; k++)
        System.out.print("  " + nodej[k]);
      System.out.println();
    }
  }
}
```

Output:

```
List of edges:
 from:   1  1  1  2  2  3  3  4  4  5  5  6
   to:   2  3  8  5  7  4  6  6  8  7  8  7
```

1.5 Random Spanning Tree

The following procedure generates a random undirected spanning tree with a given number of nodes n by means of the greedy method. The method first generates a random permutation of n objects (perm[1], perm[2], ..., perm[n]). Start with a partial tree consisting of a single node perm[1]. Suppose the current nodes in the partial tree are perm[1], perm[2],..., perm[i]. Randomly choose a node k from the partial tree, then include the node perm[i+1] and the edge (perm[i+1], k) into the partial tree. Terminate when all the n nodes are included in the partial tree.

Procedure parameters:

void randomSpanningTree (*n, seed, weighted, minweight, maxweight, nodei, nodej, weight*)

n: int;
entry: number of nodes of the tree.
Nodes of the tree are labeled from 1 to *n*.

seed: long;
entry: seed for initializing the random number generator.

weighted: boolean;
entry: *weighted* = true if the tree is weighted, false otherwise.

minweight: int;
entry: minimum weight of the edges;
if *weighted* = false then this value is ignored.

maxweight: int;
entry: maximum weight of the edges;
if *weighted* = false then this value is ignored.

nodei, nodej: int[*n*];
exit: the i-th edge is from node *nodei[i]* to node *nodej[i]*,
for i = 1,2,...,*n*–1.

weight: int[*n*];
exit: *weight[i]* is the weight of the i-th edge, for i = 1,2,...,*n*–1;
if *weighted* = false then this array is ignored.

```
public static void randomSpanningTree(int n, long seed, boolean weighted,
     int minweight, int maxweight, int nodei[], int nodej[], int weight[])
{
  int nodea,nodeb,numedges;
  int permute[] = new int[n + 1];
  Random ran = new Random(seed);

  // generate a random permutation of n objects
  randomPermutation(n,ran,permute);

  numedges = 0;
```

```
  // add n-1 random edges by the greedy method
  for (nodea=2; nodea<=n; nodea++) {
    nodeb = ran.nextInt(nodea - 1) + 1;
    numedges++;
    nodei[numedges] = permute[nodea];
    nodej[numedges] = permute[nodeb];
    if (weighted)
      weight[numedges] = (int)(minweight +
                         ran.nextDouble() * (maxweight + 1 - minweight));
  }
}
```

Example:

Generate a weighted random spanning tree of 8 nodes with edge weights in the range of [90, 99].

```
package GraphAlgorithms;

public class Test_randomSpanningTree extends Object {

  public static void main(String args[]) {

    int n = 8;
    long seed = 1;
    boolean weighted=true;
    int minweight = 90;
    int maxweight = 99;
    int nodei[] = new int[n];
    int nodej[] = new int[n];
    int weight[] = new int[n];

    GraphAlgo.randomSpanningTree(n,seed,weighted,minweight,maxweight,
                                 nodei,nodej,weight);
    System.out.println("List of edges:\n   from to  weight");
    for (int k=1; k<=n-1; k++)
      System.out.println("     " + nodei[k] + "    " + nodej[k] +
                         "    " + weight[k]);
  }
}
```

Output:

```
List of edges:
   from to  weight
     4   6    99
     2   6    93
     5   4    91
     8   4    95
     1   2    95
     3   8    96
     7   8    97
```

1.6 Random Labeled Tree

The following procedure generates a random labeled tree with a given number of nodes n. The method first constructs a Prüfer code by choosing n–2 integers $[p_1,p_2,\ldots,p_{n-2}]$ randomly among integers 1,2,...,n. Then a random labeled tree T is obtained by converting the Prüfer code [NW78]. The Prüfer code conversion starts with an empty tree T and makes use of two arrays L_1 and L_2. L_1 is initialized to contain n integers [1,2,...,n], and L_2 initially contains the n–2 integers of the Prüfer code $[p_1,p_2,\ldots, p_{n-2}]$. Let w be the smallest number in L_1 but not in L_2. Add the edge (w, p_1) to T. Delete w from L_1 and p_1 from L_2. Again let w be the smallest number in L_1 but not in L_2. Add the edge (w, p_2) to T. Delete w from L_1 and p_2 from L_2. This process stops when L_2 is empty and L_1 contains exactly two elements u and v. The tree T is completed by adding the edge (u,v).

Procedure parameters:

void randomLabeledTree (*n, seed, sol*)

n:	int;	
	entry:	number of nodes of the tree, $n > 2$. Nodes of the tree are labeled from 1 to *n*.
seed:	long;	
	entry:	seed for initializing the random number generator.
sol:	int[*n*+1];	
	exit:	the i-th edge of the tree is (i, sol[i]), i = 1,2,...,*n*–1.

```
public static void randomLabeledTree(int n, long seed, int sol[])
{
  int i,idxa,idxb,idxc,idxd,idxe,nminus2;
  int neighbor[] = new int[n + 1];
  int prufercode[] = new int[n];
  Random ran = new Random(seed);

  if (n <= 2) return;
  // select n-2 integers at random in [1,n]
```

```
  for (i=1; i<=n-2; i++)
    prufercode[i] = (int)(1 + ran.nextDouble() * n);
  // compute the tree from the Pruefer code
  for (i=1; i<=n; i++)
    sol[i] = 0;
  nminus2 = n - 2;
  for (i=1; i<=nminus2; i++) {
    idxc = prufercode[n-1-i];
    if (sol[idxc] == 0) prufercode[n-1-i] = -idxc;
    sol[idxc] = -1;
  }
  idxb = 1;
  prufercode[n-1] = n;
  idxa = 0;
  while (true) {
    if (sol[idxb] != 0) {
      idxb++;
      continue;
    }
    idxd = idxb;
    while (true) {
      idxa++;
      idxe = Math.abs(prufercode[idxa]);
      sol[idxd] = idxe;
      if (idxa == n-1) {
        for (i=1; i<=nminus2; i++)
          prufercode[i] = Math.abs(prufercode[i]);
        return;
      }
      if (prufercode[idxa] > 0) break;
      if (idxe > idxb) {
        sol[idxe] = 0;
        break;
      }
      idxd = idxe;
    }
  }
}
```

Example:

Generate a random labeled tree with 7 nodes.

```
package GraphAlgorithms;

public class Test_randomLabeledTree extends Object {

  public static void main(String args[]) {

    int n = 7;
```

```
    long seed = 1;
    int sol[] = new int[n+1];

    GraphAlgo.randomLabeledTree(n,seed,sol);
    System.out.print("List of edges:\n from: ");
    for (int k=1; k<=n-1; k++)
      System.out.print("  " + k);
    System.out.print("\n   to: ");
    for (int k=1; k<=n-1; k++)
      System.out.print("  " + sol[k]);
    System.out.println();
  }
}
```

Output:

```
List of edges:
 from:   1  2  3  4  5  6
   to:   6  3  7  3  2  7
```

1.7 Random Unlabeled Rooted Tree

The following procedure generates a random unlabeled rooted tree with a given number of nodes n. The method calculates the numbers T_1 , T_2 , ..., T_n, where T_i is the number of trees on i nodes:

$$(i-1)T_i = \sum_{v=1}^{\infty}\sum_{u=1}^{\infty} dT_{i-uv}T_v, \qquad T_1 = 1$$

A pair of positive integers u and v is chosen with a priori probability:

$$\frac{dT_{i-uv}T_v}{(i-1)T_i}$$

A random tree R_1 on n-uv nodes is chosen with probability $1/T_{n-uv}$, and a random R_2 on v nodes is chosen with probability $1/T_v$. Make u copies of R_2. A random unlabeled rooted tree on n nodes is obtained by joining the root of R_1 to the roots of each of the copies of R_2 [NW78].

Procedure parameters:

void randomUnlabeledRootedTree (*n, seed, sol*)

n:	int;	
	entry:	number of nodes of the tree, $n > 2$.
		Nodes of the tree are labeled from 1 to *n*.
seed:	long;	
	entry:	seed for initializing the random number generator.

sol: int[n+1];
exit: the i-th edge of the tree is (i, sol[i]), i = 1,2,...,n–1.

```
public static void randomUnlabeledRootedTree(int n, long seed, int sol[])
{
  int count,v,total,prod,curnum,nextlastroot,numt;
  int stackcounta,stackcountb,prob,p,q,r;
  int rightroot=0,u=0,w=0;
  int numtrees[] = new int[n + 1];
  int aux1[] = new int[n+1];
  int aux2[] = new int[n+1];
  boolean iter;
  Random ran = new Random(seed);

  // calculate numtrees[p], the number of trees on p nodes
  count = 1;
  numtrees[1] = 1;
  while (n > count) {
    total = 0;
    for (p=1; p<=count; p++) {
      q = count + 1;
      prod = numtrees[p] * p;
      for (r=1; r<=count; r++) {
        q -= p;
        if (q <= 0) break;
        total += numtrees[q] * prod;
      }
    }
    count++;
    numtrees[count] = total / (count - 1);
  }
  curnum = n;
  stackcounta = 0;
  stackcountb = 0;
  while (true) {
    if(curnum <= 2) {
      // a new tree placed in "sol", link to left neighbor
      sol[stackcountb+1] = rightroot;
      rightroot = stackcountb + 1;
      stackcountb += curnum;
      if (curnum > 1) sol[stackcountb] = stackcountb - 1;
      while (true) {
        curnum = aux2[stackcounta];
        if (curnum != 0) break;
        // stack counter is decremented as (u,v) is read
        u = aux1[stackcounta];
        stackcounta--;
        w = stackcountb - rightroot + 1;
        nextlastroot = sol[rightroot];
        numt = rightroot + (u - 1) * w - 1;
```

```
          if (u != 1) {
            // make u copies of the last tree
            for (p=rightroot; p<=numt; p++) {
              sol[p+w] = sol[p] + w;
              if((p-1)-((p-1)/w)*w == 0) sol[p+w] = nextlastroot;
            }
          }
          stackcountb = numt + w;
          if (stackcountb == n) return;
          rightroot = nextlastroot;
        }
        aux2[stackcounta] = 0;
        continue;
      }
      // choose a pair (u,v) with a priori probability
      prob = (int)((curnum - 1) * numtrees[curnum] * ran.nextDouble());
      v = 0;
      iter = true;
      while (iter) {
        v++;
        prod = v * numtrees[v];
        w = curnum;
        u = 0;
        iter = false;
        do {
          u++;
          w -= v;
          if (w < 1) {
            iter = true;
            break;
          }
          prob -= numtrees[w] * prod;
        } while (prob >= 0);
      }
      stackcounta++;
      aux1[stackcounta] = u;
      aux2[stackcounta] = v;
      curnum = w;
    }
  }
```

Example:

Generate a random unlabeled rooted tree with 5 nodes.

```
package GraphAlgorithms;

public class Test_randomUnlabeledRootedTree extends Object {

  public static void main(String args[]) {
```

```
      int n = 5;
      long seed = 1;
      int sol[] = new int[n+1];

      GraphAlgo.randomUnlabeledRootedTree(n,seed,sol);
      System.out.println("List of edges:\n   from to");
      for (int k=2; k<=n; k++)
        System.out.println("      " + k + "    " + sol[k]);
  }
}
```

Output:

```
List of edges:
   from to
     2   1
     3   2
     4   2
     5   4
```

1.8 Random Connected Graph

The following procedure [JK91] generates a random undirected connected simple graph with some given number of nodes and edges. A random spanning tree is first generated. Then random edges are added until the required number of edges is reached.

Procedure parameters:

int randomConnectedGraph (*n, m, seed, weighted, minweight, maxweight, nodei, nodej, weight*)

randomConnectedGraph: int;
- exit: the method returns the following error code:
  - 0: solution found with normal execution
  - 1: value of m is too small, should be at least n–1
  - 2: value of m is too large, should be at most $n*(n-1)/2$

n: int;
- entry: number of nodes of the graph.
  Nodes of the graph are labeled from 1 to n.

m: int;
- entry: number of edges of the graph.
  If m is less than n–1 then m will be set to n–1.
  If m is greater than the maximum number of edges in a graph then a complete graph is generated.

seed: long;
entry: seed for initializing the random number generator.
weighted: boolean;
entry: *weighted* = true if the graph is weighted, false otherwise.
minweight: int;
entry: minimum weight of the edges;
if *weighted* = false then this value is ignored.
maxweight: int;
entry: maximum weight of the edges;
if *weighted* = false then this value is ignored.
nodei, nodej: int[n];
exit: the i-th edge is from node *nodei[i]* to node *nodej[i]*,
for i = 1,2,...,n–1.
weight: int[n];
exit: *weight[i]* is the weight of the i-th edge, for i = 1,2,...,n–1;
if *weighted* = false then this array is ignored.

```
public static int randomConnectedGraph(int n, int m, long seed,
                boolean weighted, int minweight, int maxweight,
                int nodei[], int nodej[], int weight[])
{
  int maxedges,nodea,nodeb,numedges,temp;
  int permute[] = new int[n + 1];
  boolean adj[][] = new boolean[n+1][n+1];
  Random ran = new Random(seed);

  // initialize the adjacency matrix
  for (nodea=1; nodea<=n; nodea++)
    for (nodeb=1; nodeb<=n; nodeb++)
      adj[nodea][nodeb] = false;
  numedges = 0;
  // check for valid input data
  if (m < (n - 1)) return 1;
  maxedges = (n * (n - 1)) / 2;
  if (m > maxedges) return 2;

  // generate a random spanning tree by the greedy method
  randomPermutation(n,ran,permute);
  for (nodea=2; nodea<=n; nodea++) {
    nodeb = ran.nextInt(nodea - 1) + 1;
    numedges++;
    nodei[numedges] = permute[nodea];
    nodej[numedges] = permute[nodeb];
    adj[permute[nodea]][permute[nodeb]] = true;
    adj[permute[nodeb]][permute[nodea]] = true;
    if (weighted)
      weight[numedges] = (int)(minweight +
                          ran.nextDouble() * (maxweight + 1 - minweight));
  }
  // add the remaining edges randomly
```

```
    while (numedges < m) {
      nodea = ran.nextInt(n) + 1;
      nodeb = ran.nextInt(n) + 1;
      if (nodea == nodeb) continue;
      if (nodea > nodeb) {
        temp = nodea;
        nodea = nodeb;
        nodeb = temp;
      }
      if (!adj[nodea][nodeb]) {
        numedges++;
        nodei[numedges] = nodea;
        nodej[numedges] = nodeb;
        adj[nodea][nodeb] = true;
        if (weighted)
          weight[numedges] = (int)(minweight +
                              ran.nextDouble() * (maxweight + 1 - minweight));
      }
    }
    return 0;
  }
```

Example:

Generate a random undirected connected simple graph of 8 nodes and 10 edges with edge weights in the range of [90, 99].

```
package GraphAlgorithms;

public class Test_randomConnectedGraph extends Object {

  public static void main(String args[]) {

    int k;
    int n = 8;
    int m = 10;
    long seed = 1;
    boolean weighted=true;
    int minweight = 90;
    int maxweight = 99;
    int nodei[] = new int[m+1];
    int nodej[] = new int[m+1];
    int weight[] = new int[m+1];

    k = GraphAlgo.randomConnectedGraph(n,m,seed,weighted,minweight,
                                  maxweight,nodei,nodej,weight);
    if (k != 0)
      System.out.println("Invalid input data, error code = " + k);
    else {
      System.out.println("List of edges:\n from to  weight");
```

```
      for (k=1; k<=m; k++)
        System.out.println("   " + nodei[k] + "   " + nodej[k] +
                           "    " + weight[k]);
      System.out.println();
    }
  }
}
```

Output:

```
List of edges:
 from to  weight
   4   6    99
   2   6    93
   5   4    91
   8   4    95
   1   2    95
   3   8    96
   7   8    97
   2   5    98
   4   7    90
   5   7    90
```

1.9 Random Hamilton Graph

The following procedure [JK91] generates a random simple Hamilton graph with some given number of nodes and edges. The graph can be directed or undirected. The method generates a random permutation of n objects (perm[1], perm[2], ..., perm[n]). The graph is initialized with the Hamilton cycle (perm[1], perm[2]), (perm[2], perm[3]), ..., (perm[n–1], perm[n]), (perm[n], perm[1]). Then random edges are added into the graph until the required number of edges is reached.

Procedure parameters:

int randomHamiltonGraph (*n, m, seed, directed, weighted, minweight, maxweight, nodei, nodej, weight*)

randomHamiltonGraph: int;
exit: the method returns the following error code:
0: solution found with normal execution
1: value of *m* is too small, should be at least *n*
2: value of *m* is too large, should be at most $n*(n-1)/2$ for simple undirected graph, and $n*(n-1)$ for simple directed graph.

n: int;
entry: number of nodes of the graph.
Nodes of the graph are labeled from 1 to *n*.

m: int;
entry: number of edges of the graph.
If *m* is greater than the maximum number of edges in a graph then a complete graph is generated.
If *m* is less than *n* then *m* is set equal to *n*.

seed: long;
entry: seed for initializing the random number generator.

directed: boolean;
entry: *directed* = true if the graph is directed, false otherwise.

weighted: boolean;
entry: *weighted* = true if the graph is weighted, false otherwise.

minweight: int;
entry: minimum weight of the edges;
if *weighted* = false then this value is ignored.

maxweight: int;
entry: maximum weight of the edges;
if *weighted* = false then this value is ignored.

nodei, nodej: int[*m*+1];
exit: the i-th edge is from node *nodei[i]* to node *nodej[i]*, for i = 1,2,...,*m*. The Hamilton cycle is given by the first *n* elements of these two arrays.

weight: int[*m*+1];
exit: *weight[i]* is the weight of the i-th edge, for i = 1,2,...,*m*;
if *weighted* = false then this array is ignored.

```
public static int randomHamiltonGraph(int n, int m, long seed,
            boolean directed, boolean weighted, int minweight,
            int maxweight, int nodei[], int nodej[], int weight[])
{
  int k,maxedges,nodea,nodeb,numedges,temp;
  int permute[] = new int[n + 1];
  boolean adj[][] = new boolean[n+1][n+1];
  Random ran = new Random(seed);

  // initialize the adjacency matrix
  for (nodea=1; nodea<=n; nodea++)
    for (nodeb=1; nodeb<=n; nodeb++)
      adj[nodea][nodeb] = false;
  // adjust value of m if needed
  if (m < n) return 1;
  maxedges = n * (n - 1);
  if (!directed) maxedges /= 2;
  if (m > maxedges) return 2;
  numedges = 0;
  // generate a random permutation
  randomPermutation(n,ran,permute);
  // obtain the initial cycle
  for (k=1; k<=n; k++) {
    if (k == n) {
      nodea = permute[n];
```

```
      nodeb = permute[1];
    }
    else {
      nodea = permute[k];
      nodeb = permute[k + 1];
    }
    numedges++;
    nodei[numedges] = nodea;
    nodej[numedges] = nodeb;
    adj[nodea][nodeb] = true;
    if (!directed) adj[nodeb][nodea] = true;
    if (weighted)
      weight[numedges] = (int)(minweight +
                           ran.nextDouble() * (maxweight + 1 - minweight));
  }
  // add the remaining edges randomly
  while (numedges < m) {
    nodea = ran.nextInt(n) + 1;
    nodeb = ran.nextInt(n) + 1;
    if (nodea == nodeb) continue;
    if ((nodea > nodeb) && (!directed)) {
      temp = nodea;
      nodea = nodeb;
      nodeb = temp;
    }
    if (!adj[nodea][nodeb]) {
      numedges++;
      nodei[numedges] = nodea;
      nodej[numedges] = nodeb;
      adj[nodea][nodeb] = true;
      if (weighted)
        weight[numedges] = (int)(minweight +
                           ran.nextDouble() * (maxweight + 1 - minweight));
    }
  }
  return 0;
}
```

Example:

Generate a random simple Hamilton graph of 7 nodes and 10 edges with edge weights in the range of [90, 99].

```
package GraphAlgorithms;

public class Test_randomHamiltonGraph extends Object {

  public static void main(String args[]) {

    int k;
```

```
        int n = 7;
        int m = 10;
        long seed = 1;
        boolean weighted=true;
        boolean directed=true;
        int minweight = 90;
        int maxweight = 99;
        int nodei[] = new int[m+1];
        int nodej[] = new int[m+1];
        int weight[] = new int[m+1];

        k = GraphAlgo.randomHamiltonGraph(n,m,seed,directed,weighted,minweight,
                                          maxweight,nodei,nodej,weight);
        if (k != 0)
          System.out.println("Invalid input data, error code = " + k);
        else {
          System.out.println("List of edges:\n from to  weight");
          for (k=1; k<=m; k++)
            System.out.println("    " + nodei[k] + "    " + nodej[k] +
                               "     " + weight[k]);
        }
    }
}
```

Output:

```
List of edges:
 from to  weight
   6   4    99
   4   2    99
   2   5    99
   5   7    93
   7   1    93
   1   3    92
   3   6    95
   6   3    97
   3   4    91
   4   7    91
```

1.10 Random Maximum Flow Network

The following procedure [JK91] generates a random simple weighted directed graph of n nodes in which node 1 (the source) has no incoming edges and node n (the sink) has no outgoing edges. The procedure first attempts to generate random paths until either the required number of edges is reached or every node is on a directed path from the source to the sink. If each node has already been included in some directed path from the source to the sink then additional edges (p,q), where 0<p<n and 1<q≤n, are generated randomly until the required number of edges is reached.

Procedure parameters:

void randomMaximumFlowNetwork (*n, m, seed, minweight,maxweight, nodei, nodej, weight*)

n: int;
entry: number of nodes of the network.
Nodes of the graph are labeled from 1 to *n*.
Node 1 is the source and node *n* is the sink.

m: int;
entry: number of required edges of the directed graph.

seed: long;
entry: seed for initializing the random number generator.

minweight: int;
entry: minimum weight of the edges.

maxweight: int;
entry: maximum weight of the edges.

nodei, nodej: int[*m*+1];
exit: *nodei[0]* returns the actual number of edges generated.
If $m \le (n * n - 3 * n + 3)$ then *nodei[0]* = m, otherwise *nodei[0]* = $(n * n - 3 * n + 3)$.
The i-th edge is from node *nodei[i]* to node *nodej[i]*, for i = 1,2,...,*nodei[0]*.

weight: int[*m*+1];
exit: *weight[i]* is the weight of the i-th edge, for i=1,2,...,*nodei[0]*.

```
public static void randomMaximumFlowNetwork(int n, int m, long seed,
                          int minweight, int maxweight, int nodei[],
                          int nodej[], int weight[])
{
  int i,maxedges,nodea,nodeb,numedges,source,sink;
  boolean adj[][] = new boolean[n+1][n+1];
  boolean marked[] = new boolean[n+1];
  boolean more;
  Random ran = new Random(seed);

  if ((n <= 1) || (m < 1)) return;
  // initialize the adjacency matrix
  for (nodea=1; nodea<=n; nodea++)
    for (nodeb=1; nodeb<=n; nodeb++)
      adj[nodea][nodeb] = false;
  // check for valid input data
  maxedges = n * n - 3 * n + 3;
  if (m > maxedges) m = maxedges;
  nodei[0] = m;

  // node 1 is the source and node n is the sink
```

```
source = 1;
sink = n;
numedges = 0;
// initially every node is not on some path from source to sink
for (i=1; i<=n; i++)
  marked[i] = false;
// include each node on some path from source to sink */
marked[source] = true;
do {
  // choose an edge from source to some node not yet included
  do {
    // generate a random integer in interval [2,n]
    nodeb = (int)(2 + ran.nextDouble() * (n-1));
  } while (marked[nodeb]);
  marked[nodeb] = true;
  // add the edge from source to nodeb
  adj[1][nodeb] = true;
  numedges++;
  weight[numedges] = (int)(minweight +
                     ran.nextDouble() * (maxweight + 1 - minweight));
  nodei[numedges] = 1;
  nodej[numedges] = nodeb;
  if (numedges == m) return;
  // add an edge from current node to a node other than the sink
  if (nodeb != sink) {
    nodea = nodeb;
    marked[sink] = false;
    while (true) {
      do {
        // generate a random integer in interval [2,n]
        nodeb = (int)(2 + ran.nextDouble() * (n-1));
      } while (marked[nodeb]);
      marked[nodeb] = true;
      if (nodeb == sink) break;
      // add an edge from nodea to nodeb
      adj[nodea][nodeb] = true;
      numedges++;
      weight[numedges] = (int)(minweight +
                         ran.nextDouble() * (maxweight + 1 - minweight));
      nodei[numedges] = nodea;
      nodej[numedges] = nodeb;
      if (numedges == m) return;
      nodea = nodeb;
    }
    // add an edge from nodea to sink
    adj[nodea][sink] = true;
    numedges++;
    weight[numedges] = (int)(minweight +
                       ran.nextDouble() * (maxweight + 1 - minweight));
    nodei[numedges] = nodea;
```

```
      nodej[numedges] = sink;
      if (numedges == m) return;
    }
    more = false;
    for (i=1; i<n; i++)
      if (!marked[i]) {
        more = true;
        break;
      }
  } while (more);
  // add additional edges if needed
  while (numedges < m) {
    // generate a random integer in interval [1,n-1]
    nodea = (int)(1 + ran.nextDouble() * (n-1));
    // generate a random integer in interval [2,n]
    nodeb = (int)(2 + ran.nextDouble() * (n-1));
    if (!adj[nodea][nodeb] && (nodea != nodeb)) {
      // add an edge from nodea to nodeb
      adj[nodea][nodeb] = true;
      numedges++;
      weight[numedges] = (int)(minweight +
                        ran.nextDouble() * (maxweight + 1 - minweight));
      nodei[numedges] = nodea;
      nodej[numedges] = nodeb;
    }
  }
}
```

Example:

Generate a random maximum flow network of 8 nodes and 10 edges with edge weights in the range of [90, 99]. Node 1 is the source and node 8 is the sink.

```
package GraphAlgorithms;

public class Test_randomMaximumFlowNetwork extends Object {

  public static void main(String args[]) {

    int n = 8;
    int m = 10;
    long seed = 1;
    int minweight = 90;
    int maxweight = 99;
    int nodei[] = new int[m+1];
    int nodej[] = new int[m+1];
    int weight[] = new int[m+1];

    GraphAlgo.randomMaximumFlowNetwork(n,m,seed,minweight,maxweight,
                                       nodei,nodej,weight);
```

```
    System.out.println("List of edges:\n from to  weight");
    for (int k=1; k<=nodei[0]; k++)
      System.out.println("    " + nodei[k] + "    " + nodej[k] +
                         "     " + weight[k]);
  }
}
```

Output:

```
List of edges:
 from to  weight
   1   7    94
   7   3    93
   3   8    90
   1   4    93
   4   5    91
   5   6    91
   6   2    96
   2   8    92
   4   3    98
   1   6    96
```

1.11 Random Isomorphic Graphs

The following procedure generates a pair of random isomorphic graphs with some given number of nodes and edges. The graph can be simple or nonsimple, directed or undirected. The method generates a random graph first, then a random permutation of n objects (perm[1], perm[2], ..., perm[n]). The second isomorphic graph is obtained by renaming the vertices of the first random graph by the random permutation. The node i of the first random graph corresponds to the node perm[i] in the second graph.

Procedure parameters:

int randomIsomorphicGraphs (*n, m, seed, simple, directed, firsti, firstj, secondi, secondj, map*)

randomIsomorphicGraph: int;
- exit: the method returns the following error code:
  - 0: solution found with normal execution
  - 1: value of m is too large, should be at most $n*(n-1)/2$ for simple undirected graph, and $n*(n-1)$ for simple directed graph.

n: int;
- entry: number of nodes of each graph. Nodes of each graph are labeled from 1 to n.

m: int;
entry: number of edges of each graph.
If *m* is greater than the maximum number of edges in a graph then a complete graph is generated.

seed: long;
entry: seed for initializing the random number generator.

simple: boolean;
entry: *simple* = true if the graphs are simple, false otherwise.

directed: boolean;
entry: *directed* = true if the graphs are directed, false otherwise.

firsti, firstj: int[*m*+1];
exit: the k-th edge of the first graph is from node *firsti[k]* to node *firstj[k]*, for k = 1,2,...,*m*.

secondi, secondj: int[*m*+1];
exit: the k-th edge of the second graph is from node *secondi[k]* to node *secondj[k]*, for k = 1,2,...,*m*.

map: int[*n*+1];
exit: in the graph isomorphism, node i of the first graph is renamed to node *map[i]* in the second graph, for i=1,2,...,*n*.

```
public static int randomIsomorphicGraphs(int n, int m, long seed,
       boolean simple, boolean directed, int firsti[], int firstj[],
       int secondi[], int secondj[], int map[])
{
  int k;
  Random ran = new Random(seed);

  // generate a random graph
  k = randomGraph(n,m,seed,simple,directed,false,false,0,0,firsti,firstj,map);
  if (k != 0) return k;
  // generate a random permutation
  randomPermutation(n,ran,map);
  // rename the vertices to obtain the isomorphic graph
  for (int i=1; i<=m; i++) {
    secondi[i] = map[firsti[i]];
    secondj[i] = map[firstj[i]];
  }
  return k;
}
```

Example:

Generate a pair of random isomorphic graphs with 5 nodes and 7 edges.

```
package GraphAlgorithms;

public class Test_randomIsomorphicGraphs extends Object {

  public static void main(String args[]) {
```

```
    int k;
    int n = 5;
    int m = 7;
    long seed = 1;
    boolean simple=true, directed=false;
    int map[] = new int[n+1];
    int firsti[] = new int[m+1];
    int firstj[] = new int[m+1];
    int secondi[] = new int[m+1];
    int secondj[] = new int[m+1];

    k = GraphAlgo.randomIsomorphicGraphs(n,m,seed,simple,directed,
                              firsti,firstj,secondi,secondj,map);
    if (k != 0)
      System.out.println("Invalid input data, error code = " + k);
    else {
      System.out.println("List of edges:\n  First Graph    Second Graph" +
                         "\n   from to          from to ");
      for (k=1; k<=m; k++)
        System.out.println("     " + firsti[k] + "    " + firstj[k] +
                           "           " + secondi[k] + "    " + secondj[k]);
      System.out.println("\n  Node mapping:\n   First Graph   Second Graph");
      for (k=1; k<=n; k++)
        System.out.println("        " + k + "            " + map[k]);
    }
  }
}
```

Output:

```
List of edges:
  First Graph    Second Graph
   from to          from to
     1    4           4    1
     3    4           2    1
     2    5           3    5
     4    5           1    5
     3    5           2    5
     2    3           3    2
     1    5           4    5

  Node mapping:
   First Graph   Second Graph
        1            4
        2            3
        3            2
        4            1
        5            5
```

1.12 Random Isomorphic Regular Graphs

The following procedure generates a pair of random isomorphic simple undirected regular graphs with some given number of nodes and the degree of each node. The method generates a random graph first, then a random permutation of n objects (perm[1], perm[2], ..., perm[n]). The second isomorphic graph is obtained by renaming the vertices of the first random graph by the random permutation. The node i of the first random graph corresponds to the node perm[i] in the second graph.

Procedure parameters:

int randomIsomorphicRegularGraphs (*n, degree, seed, firsti, firstj,secondi, secondj, map*)

randomIsomorphicRegularGraphs: int;
exit: the method returns the following error code:
0: solution found with normal execution
1: invalid input, if *degree* is odd then *n* must be even
2: value of *n* should be greater than *degree*

n: int;
entry: number of nodes of each graph.
Nodes of each graph are labeled from 1 to *n*.

degree: int;
entry: required degree of each node.
If the value of *degree* is odd then the value *n* must be even.

seed: long;
entry: seed for initializing the random number generator.

firsti, firstj: int[(*n***degree*)/2 + 1];
exit: the k-th edge of the first graph is from node *firsti[k]* to node *firstj[k]*, for k = 1, 2, ..., (*n***degree*)/2.

secondi, secondj: int[(*n***degree*)/2 + 1];
exit: the k-th edge of the second graph is from node *secondi[k]* to node *secondj[k]*, for k = 1, 2, ..., (*n***degree*)/2.

map: int[*n*+1];
exit: in the graph isomorphism, node i of the first graph is renamed to node *map[i]* in the second graph, for i = 1,2,...,*n*.

```
public static int randomIsomorphicRegularGraphs(int n, int degree, long seed,
          int firsti[], int firstj[], int secondi[], int secondj[], int map[])
{
  int k;
  Random ran = new Random(seed);

  // generate a random regular graph
  k = randomRegularGraph(n,degree,seed,firsti,firstj);
  if (k != 0) return k;
  // generate a random permutation
  randomPermutation(n,ran,map);
```

```
    // rename the vertices to obtain the isomorphic graph
    for (int i=1; i<=(n*degree)/2; i++) {
      secondi[i] = map[firsti[i]];
      secondj[i] = map[firstj[i]];
    }
    return k;
}
```

Example:

Generate a pair of random isomorphic simple undirected regular graphs with 6 nodes of degree 3.

```
package GraphAlgorithms;

public class Test_randomIsomorphicRegularGraphs extends Object {

  public static void main(String args[]) {

    int k;
    int n = 6;
    int degree = 3;
    long seed = 1;
    int m = (n * degree) / 2;
    int map[] = new int[n+1];
    int firsti[] = new int[m+1];
    int firstj[] = new int[m+1];
    int secondi[] = new int[m+1];
    int secondj[] = new int[m+1];

    k = GraphAlgo.randomIsomorphicRegularGraphs(n,degree,seed,
                        firsti,firstj,secondi,secondj,map);
    if (k != 0)
      System.out.println("Invalid input data, error code = " + k);
    else {
      System.out.println("List of edges:\n  First Graph    Second Graph" +
                        "\n   from to         from to ");
      for (k=1; k<=m; k++)
        System.out.println("      " + firsti[k] + "   " + firstj[k] +
                        "           " + secondi[k] + "   " + secondj[k]);
      System.out.println("\n  Node mapping:\n   First Graph   Second Graph");
      for (k=1; k<=n; k++)
        System.out.println("        " + k + "              " + map[k]);
    }
  }
}
```

Output:

```
List of edges:
  First Graph     Second Graph
   from to         from to
     1   3           5   3
     1   4           5   2
     1   6           5   1
     2   3           4   3
     2   4           4   2
     2   5           4   6
     3   6           3   1
     4   5           2   6
     5   6           6   1


  Node mapping:
   First Graph    Second Graph
        1             5
        2             4
        3             3
        4             2
        5             6
        6             1
```

2. Connectivity

2.1 Maximum Connectivity

Given two positive integers n and k, the following procedure [H62] constructs a k-connected simple undirected graph $G(k,n)$ on n nodes with the least number of edges. It is known that $G(k,n)$ has exactly $\lceil (n*k)/2 \rceil$ edges, where $\lceil x \rceil$ is the smallest integer greater than or equal to x. Note that $G(1,n)$ is a spanning tree. It is assumed that $k \geq 2$. Let the nodes of the graph be integers $0,1,2,\ldots,n-1$.

Case 1: (k is even)

Let $k = 2t$. The graph $G(2t,n)$ is constructed as follows. Draw an n-gon, that is, add the edges (0,1), (1,2), (2,3), ..., $(n-2,n-1)$, $(n-1,0)$. Then join nodes i and j if and only if $|i-j| \equiv p$ (mode n), where $2 \leq p \leq t$.

Case 2: (k is odd, n is even)

Let $k = 2t+1$. The graph $G(2t+1,n)$ is constructed by first drawing $G(2t,n)$, and then joining node i to node $i+(n/2)$, for $0 \leq i < n/2$.

Case 1: (k is odd, n is odd)

Let $k = 2t+1$. The graph $G(2t+1,n)$ is constructed by first drawing $G(2t,n)$, and then joining:

node 0 to node $(n-1)/2$,
node 0 to node $(n+1)/2$,
node i to node $i+(n+1)/2$, for $1 \leq i < (n-1)/2$.

Procedure parameters:

void maximumConnectivity (n, k, *nodei*, *nodej*)

n: int;
entry: the number of nodes of the undirected graph, labeled from 1 to n.

k: int;
entry: the required graph is k-connected, $k \geq 2$.

nodei, nodej: int[$\lceil (n*k)/2 \rceil$ + 1];
exit: *nodei[p]* and *nodej[p]* are the end nodes of the p-th edge in the k-connected graph, $p=1,2,\ldots,\lceil (n*k)/2 \rceil$.

```
public static void maximumConnectivity(int n, int k, int nodei[], int nodej[])
{
  int edges,halfk,halfn,i,j,nminus1,p,q,r;
  boolean evenk,evenn,join;

  // make an n-gon
  edges = 0;
  nminus1  = n - 1;
  halfk = k / 2;
  halfn = n / 2;
  for (i=1; i<=nminus1; i++) {
    edges++;
    nodei[edges] = i;
    nodej[edges] = i + 1;
  }
```

```
  edges++;
  nodei[edges] = n;
  nodej[edges] = 1;
  if (k == 2) return;
  evenk = (k == 2 * halfk) ? true : false;
  for (i=1; i<=nminus1; i++) {
    p = i + 1;
    for (j=p; j<=n; j++) {
      join = false;
      q = j - i;
      for (r=2; r<=halfk; r++)
        if (((r - ((r/n)*n)) == q) || (q + r == n)) join = true;
      if (join) {
        edges++;
        nodei[edges] = i;
        nodej[edges] = j;
      }
    }
  }
  // if k is even then finish
  if (evenk) return;
  evenn = (n == 2 * halfn) ? true : false;
  if (evenn) {
    // k is odd, n is even
    for (i=1; i<=halfn; i++) {
      edges++;
      nodei[edges] = i;
      nodej[edges] = i + halfn;
    }
  }
  else {
    // k is odd, n is odd
    p = (n + 1) / 2;
    q = (n - 1) / 2;
    for (i=2; i<=q; i++) {
      edges++;
      nodei[edges] = i;
      nodej[edges] = i + p;
    }
    edges++;
    nodei[edges] = 1;
    nodej[edges] = q + 1;
    edges++;
    nodei[edges] = 1;
    nodej[edges] = p + 1;
  }
}
```

Example:

Construct a 4-connected simple undirected graph on 8 nodes with the least number of edges.

```
package GraphAlgorithms;

public class Test_maximumConnectivity extends Object {

  public static void main(String args[]) {

    int n = 8;
    int k = 4;
    int nk2 = (n * k) / 2;
    int nodei[] = new int[nk2+1];
    int nodej[] = new int[nk2+1];

    GraphAlgo.maximumConnectivity(n,k,nodei,nodej);
    System.out.println("List of edges:");
    for (int p=1; p<=nk2; p++)
      System.out.print("  " + nodei[p]);
    System.out.println();
    for (int p=1; p<=nk2; p++)
      System.out.print("  " + nodej[p]);
    System.out.println();
  }
}
```

Output:

```
List of edges:
  1  2  3  4  5  6  7  8  1  1  2  2  3  4  5  6
  2  3  4  5  6  7  8  1  3  7  4  8  5  6  7  8
```

2.2 Depth-First Search

The following procedure [G85] visits the nodes of an undirected graph systematically by a depth-first search. Starting by choosing any node p, the method visits any node q adjacent to p that has not been visited. The node q is put on the stack. After examining p, the top element r is removed from the stack. Then r is processed the same way as before. When the stack is empty all the nodes of the current component will have been visited. Choose any node from the next component of the graph and repeat the same procedure. The algorithm terminates when all nodes are visited. The running time is $O(\max(n,m))$, where n is the number of nodes and m is the number of edges of the graph.

Procedure parameters:

void depthFirstSearch (*n, m, nodei, nodej, parent, sequence*)

n: int;
entry: number of nodes of the undirected graph,
nodes of the graph are labeled from 1 to *n*.

m: int;
entry: number of edges of the undirected graph.

nodei, nodej: int[*m*+1];
entry: *nodei[p]* and *nodej[p]* are the end nodes of the p-th edge in the graph, p=1,2,...,*m*.

parent: int[*n*+1];
exit: *parent[i]* is the previous node which was visited just before node i; *parent[i]=0* if node i is the first node being visited in the component, for i=1,2,...,*n*.

sequence: int[*n*+1];
exit: *sequence[i]* is the order in which node i was visited in the search, for i=1,2,...,*n*.

```
public static void depthFirstSearch(int n, int m, int nodei[], int nodej[],
                                    int parent[], int sequence[])
{
  int i,j,k,counter,stackindex,p,q,u,v;
  int stack[] = new int[n+1];
  int firstedges[] = new int[n+2];
  int endnode[] = new int[m+1];
  boolean mark[] = new boolean[m+1];
  boolean skip,found;

  // set up the forward star representation of the graph
  for (j=1; j<=m; j++)
    mark[j] = true;
  firstedges[1] = 0;
  k = 0;
  for (i=1; i<=n; i++) {
    for (j=1; j<=m; j++)
      if (mark[j]) {
        if (nodei[j] == i) {
          k++;
          endnode[k] = nodej[j];
          mark[j] = false;
        }
        else {
          if (nodej[j] == i) {
            k++;
            endnode[k] = nodei[j];
            mark[j] = false;
          }
        }
```

```
    }
    firstedges[i+1] = k;
  }
  for (i=1; i<=n; i++) {
    sequence[i] = 0;
    parent[i] = 0;
    stack[i] = 0;
  }
  counter = 0;
  p = 1;
  stackindex = 0;
  // process descendents of node p
  while (true) {
    skip = false;
    counter++;
    parent[p] = 0;
    sequence[p] = counter;
    stackindex++;
    stack[stackindex] = p;
    while (true) {
      skip = false;
      q = 0;
      while (true) {
        q++;
        if (q <= n) {
          // check if p and q are adjacent
          if (p < q) {
            u = p;
            v = q;
          }
          else {
            u = q;
            v = p;
          }
          found = false;
          for (k=firstedges[u]+1; k<=firstedges[u+1]; k++)
            if (endnode[k] == v) {
              // u and v are adjacent
              found = true;
              break;
            }
          if (found && sequence[q] == 0) {
            stackindex++;
            stack[stackindex] = q;
            parent[q] = p;
            counter++;
            sequence[q] = counter;
            p = q;
            if (counter == n) return;
            break;
```

```
          }
        }
        else {
          // back up
          stackindex--;
          if (stackindex > 0) {
            q = p;
            p = stack[stackindex];
          }
          else {
            skip = true;
            break;
          }
        }
      }
      if (skip) break;
    }
    // process the next component
    stackindex = 0;
    skip = false;
    for (k=1; k<=n; k++)
      if (sequence[k] == 0) {
        p = k;
        skip = true;
        break;
      }
    if (!skip) break;
  }
}
```

Example:

Apply a depth-first search to the following graph.

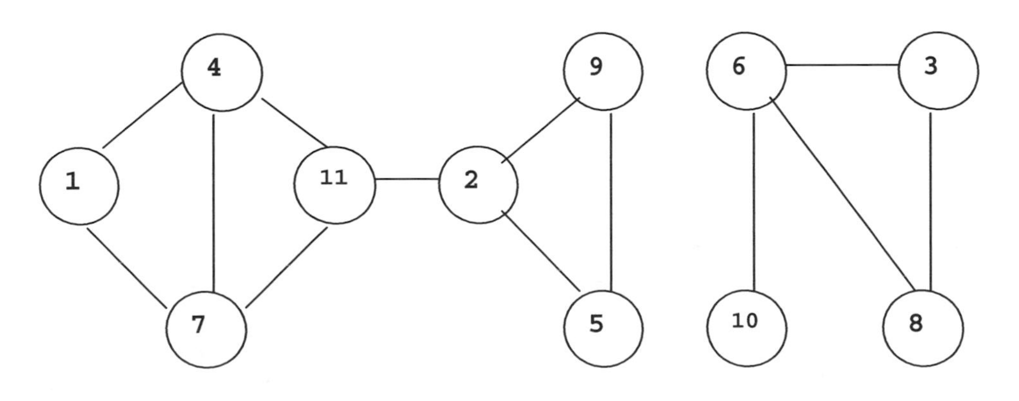

```
package GraphAlgorithms;

public class Test_depthFirstSearch extends Object {

  public static void main(String args[]) {
    int n=11;
    int m=13;
    int nodei[] = {0,7,1,6,3, 4,2,11,8,9, 6,4, 2,5};
    int nodej[] = {0,4,7,8,6,11,5, 7,3,2,10,1,11,9};
    int parent[] = new int[n+1];
    int sequence[] = new int[n+1];

    GraphAlgo.depthFirstSearch(n,m,nodei,nodej,parent,sequence);
    System.out.printf("Depth-first search results:\n   Node:");
    for (int i=1; i<=n; i++)
      System.out.printf("%4d",i);
    System.out.printf("\n Parent:");
    for (int i=1; i<=n; i++)
      System.out.printf("%4d",parent[i]);
    System.out.printf("\n  Order:");
    for (int i=1; i<=n; i++)
      System.out.printf("%4d",sequence[i]);
    System.out.println();
  }
}
```

Output:

```
Depth-first search results:
   Node:   1   2   3   4   5   6   7   8   9  10  11
 Parent:   0  11   0   1   2   3   4   6   5   6   7
  Order:   1   5   8   2   6   9   3  10   7  11   4
```

2.3 Breadth-First Search

The following procedure [G85] visits the nodes of an undirected graph systematically by a breadth-first search. From a current node p, visit every node adjacent to p, and the nodes are added to a queue. When all neighbors of p are visited, remove an element q from the queue. Process node q in the same way by visiting all neighbors of q that have not been visited. When the queue is empty all the nodes in the current components will have been visited. Choose any node from the next component of the graph and repeat the same procedure. The algorithm terminates when all nodes are visited. The running time is $O(\max(n,m))$, where n is the number of nodes and m is the number of edges of the graph.

Procedure parameters:

void breadthFirstSearch (*n, m, nodei, nodej, parent, sequence*)

n: int;
entry: number of nodes of the undirected graph, nodes of the graph are labeled from 1 to *n*.

m: int;
entry: number of edges of the undirected graph.

nodei, nodej: int[*m*+1];
entry: *nodei[p]* and *nodej[p]* are the end nodes of the p-th edge in the graph, p=1,2,...,*m*.

parent: int[*n*+1];
exit: *parent[i]* is the previous node which was visited just before node i; *parent[i]=0* if node i is the first node being visited in the component, for i=1,2,...,*n*.

sequence: int[*n*+1];
exit: *sequence[i]* is the order in which node i was visited in the search, for i=1,2,...,*n*.

```
public static void breadthFirstSearch(int n, int m, int nodei[], int nodej[],
                                      int parent[], int sequence[])
{
  int i,j,k,enqueue,dequeue,queuelength,p,q,u,v;
  int queue[] = new int[n+1];
  int firstedges[] = new int[n+2];
  int endnode[] = new int[m+1];
  boolean mark[] = new boolean[m+1];
  boolean iterate,found;

  // set up the forward star representation of the graph
  for (j=1; j<=m; j++)
    mark[j] = true;
  firstedges[1] = 0;
  k = 0;
  for (i=1; i<=n; i++) {
    for (j=1; j<=m; j++)
      if (mark[j]) {
        if (nodei[j] == i) {
          k++;
          endnode[k] = nodej[j];
          mark[j] = false;
        }
        else {
          if (nodej[j] == i) {
            k++;
            endnode[k] = nodei[j];
            mark[j] = false;
          }
        }
```

```
    }
    firstedges[i+1] = k;
  }
  for (i=1; i<=n; i++) {
    sequence[i] = 0;
    parent[i] = 0;
  }
  k = 0;
  p = 1;
  enqueue = 1;
  dequeue = 1;
  queuelength = enqueue;
  queue[enqueue] = p;
  k++;
  sequence[p] = k;
  parent[p] = 0;
  iterate = true;
  // store all descendants
  while (iterate) {
    for (q=1; q<=n; q++) {
      // check if p and q are adjacent
      if (p < q) {
        u = p;
        v = q;
      }
      else {
        u = q;
        v = p;
      }
      found = false;
      for (i=firstedges[u]+1; i<=firstedges[u+1]; i++)
        if (endnode[i] == v) {
          // p and q are adjacent
          found = true;
          break;
        }
      if (found && sequence[q] == 0) {
        enqueue++;
        if (n < enqueue) enqueue = 1;
        queue[enqueue] = q;
        k++;
        parent[q] = p;
        sequence[q] = k;
      }
    }
    // process all nodes of the same height
    if (enqueue >= dequeue) {
      if (dequeue == queuelength) {
        queuelength = enqueue;
      }
```

```
      p = queue[dequeue];
      dequeue++;
      if (n < dequeue) dequeue = 1;
      iterate = true;
      // process other components
    }
    else {
      iterate = false;
      for (i=1; i<=n; i++)
        if (sequence[i] == 0) {
          dequeue = 1;
          enqueue = 1;
          queue[enqueue] = i;
          queuelength = 1;
          k++;
          sequence[i] = k;
          parent[i] = 0;
          p = i;
          iterate = true;
          break;
        }
    }
  }
}
```

Example:

Apply a breadth-first search to the following graph.

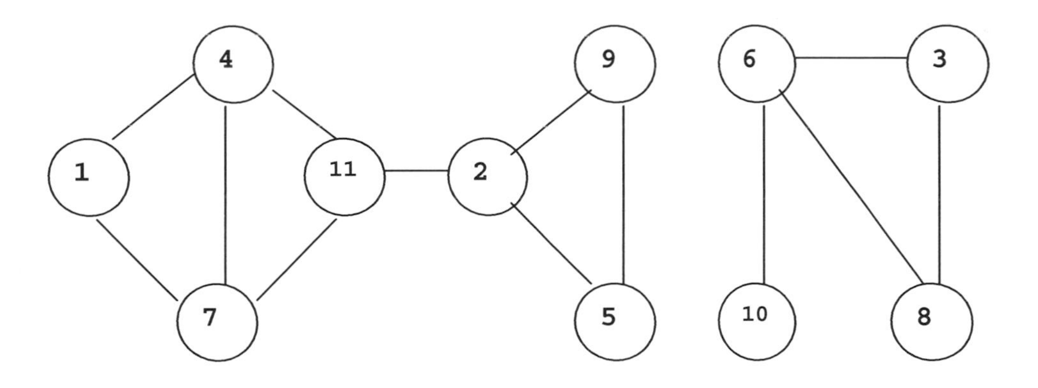


```
package GraphAlgorithms;

public class Test_breadthFirstSearch extends Object {

  public static void main(String args[]) {
    int n=11;
    int m=13;
```

```
    int nodei[] = {0,7,1,6,3, 4,2,11,8,9, 6,4, 2,5};
    int nodej[] = {0,4,7,8,6,11,5, 7,3,2,10,1,11,9};
    int parent[] = new int[n+1];
    int sequence[] = new int[n+1];

    GraphAlgo.breadthFirstSearch(n,m,nodei,nodej,parent,sequence);
    System.out.printf("Breadth-first search results:\n   Node:");
    for (int i=1; i<=n; i++)
      System.out.printf("%4d",i);
    System.out.printf("\n Parent:");
    for (int i=1; i<=n; i++)
      System.out.printf("%4d",parent[i]);
    System.out.printf("\n  Order:");
    for (int i=1; i<=n; i++)
      System.out.printf("%4d",sequence[i]);
    System.out.println();
  }
}
```

Output:

```
Breadth-first search results:
   Node:   1   2   3   4   5   6   7   8   9  10  11
 Parent:   0  11   0   1   2   3   1   3   2   6   4
  Order:   1   5   8   2   6   9   3  10   7  11   4
```

2.4 Connected Graph Testing

The following linear time procedure [BD80] checks whether a given undirected graph is connected by scanning the neighbors of all nodes.

Procedure parameters:

boolean connected (*n, m, nodei, nodej*)

connected: boolean;
  exit: *connected* = *true* if the input graph is connected, *false* otherwise.
n: int;
  entry: the number of nodes of the undirected graph, labeled from 1 to *n*.
m: int;
  entry: the number of edges in the graph.
nodei, nodej: int[*m*+1];
  entry: *nodei[i]* and *nodej[i]* are the end nodes of the i-th edge in the graph, i=1,2,...,*m*.

```
public static boolean connected(int n, int m, int nodei[], int nodej[])
{
  int i,j,k,r,connect;
  int neighbor[] = new int[m + m + 1];
  int degree[] = new int[n + 1];
  int index[] = new int[n + 2];
  int aux1[] = new int[n + 1];
  int aux2[] = new int[n + 1];

  for (i=1; i<=n; i++)
    degree[i] = 0;
  for (j=1; j<=m; j++) {
    degree[nodei[j]]++;
    degree[nodej[j]]++;
  }
  index[1] = 1;
  for (i=1; i<=n; i++) {
    index[i+1] = index[i] + degree[i];
    degree[i] = 0;
  }
  for (j=1; j<=m; j++) {
    neighbor[index[nodei[j]] + degree[nodei[j]]] = nodej[j];
    degree[nodei[j]]++;
    neighbor[index[nodej[j]] + degree[nodej[j]]] = nodei[j];
    degree[nodej[j]]++;
  }
  for (i=2; i<=n; i++)
    aux1[i] = 1;
  aux1[1] = 0;
  connect = 1;
  aux2[1] = 1;
  k = 1;
  while (true) {
    i = aux2[k];
    k--;
    for (j=index[i]; j<=index[i+1]-1; j++) {
      r = neighbor[j];
      if (aux1[r] != 0) {
        connect++;
        if (connect == n) {
          connect /= n;
          if (connect == 1) return true;
          return false;
        }
        aux1[r] = 0;
        k++;
        aux2[k] = r;
      }
    }
    if (k == 0) {
```

```
      connect /= n;
      if (connect == 1) return true;
      return false;
    }
  }
}
```

Example:

Test if the following graph is connected:

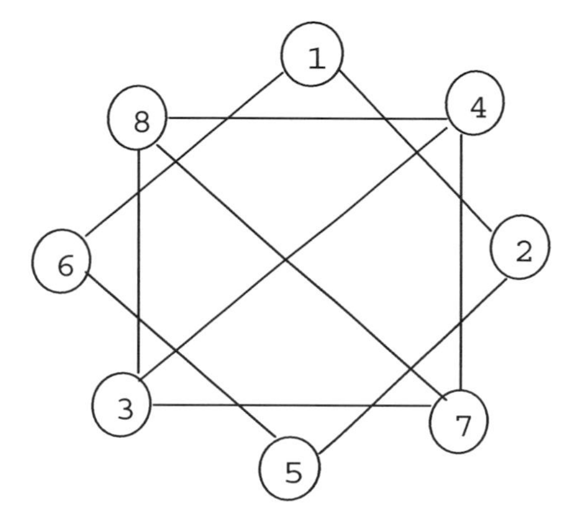


```
package GraphAlgorithms;

public class Test_connected extends Object {

  public static void main(String args[]) {

    int n = 8;
    int m = 10;
    int nodei[] = {0, 8, 3, 5, 1, 3, 7, 2, 6, 7, 4};
    int nodej[] = {0, 4, 7, 2, 6, 8, 4, 1, 5, 8, 3};

    if (GraphAlgo.connected(n, m, nodei, nodej))
      System.out.println("The input graph is connected.");
    else
      System.out.println("The input graph is not connected.");
  }
}
```

Output:

```
The input graph is not connected.
```

2.5 Connected Components

A *connected component* is a maximal connected subgraph. The following procedure [NW75] performs a linear scan of the edges and identifies the different connected components of a given undirected graph G of n nodes and m edges with $O(n \log n) + O(m)$ operations.

Procedure parameters:

void connectedComponents (*n, m, nodei, nodej, component*)

n: int;
entry: the number of nodes of the undirected graph, labeled from 1 to *n*.

m: int;
entry: the number of edges of the graph.

nodei, nodej: int[*m*+1];
entry: *nodei[i]* and *nodej[i]* are the end nodes of the i-th edge in the graph, i=1,2,...,*m*.

component: int[*n*+1];
exit: *component[0]* is the total number of components of the graph. *component[i]* is the component to which node i belongs, i=1,2,...,*n*.

```
public static void connectedComponents(int n, int m, int nodei[], int nodej[],
                                        int component[])
{
  int edges,i,j,numcomp,p,q,r,typea,typeb,typec,tracka,trackb;
  int compkey,key1,key2,key3,nodeu,nodev;
  int numnodes[] = new int[n+1];
  int aux[] = new int[n+1];
  int index[] = new int[3];

  typec=0;
  index[1] = 1;
  index[2] = 2;
  q = 2;
  for (i=1; i<=n; i++) {
    component[i] = -i;
    numnodes[i] = 1;
    aux[i] = 0;
  }
  j = 1;
  edges = m;
  do {
    nodeu = nodei[j];
    nodev = nodej[j];
    key1 = component[nodeu];
    if (key1 < 0) key1 = nodeu;
    key2 = component[nodev];
    if (key2 < 0) key2 = nodev;
```

```
    if (key1 == key2) {
      if(j >= edges) {
        edges--;
        break;
      }
      nodei[j] = nodei[edges];
      nodej[j] = nodej[edges];
      nodei[edges] = nodeu;
      nodej[edges] = nodev;
      edges--;
    }
    else {
      if (numnodes[key1] >= numnodes[key2]) {
        key3 = key1;
        key1 = key2;
        key2 = key3;
        typec = -component[key2];
      }
      else
        typec = Math.abs(component[key2]);
      aux[typec] = key1;
      component[key2] = component[key1];
      i = key1;
      do {
        component[i] = key2;
        i = aux[i];
      } while (i != 0);
      numnodes[key2] += numnodes[key1];
      numnodes[key1] = 0;
      j++;
      if (j > edges || j > n) break;
    }
  } while (true);
  numcomp = 0;
  for (i=1; i<=n; i++)
    if (numnodes[i] != 0) {
      numcomp++;
      numnodes[numcomp] = numnodes[i];
      aux[i] = numcomp;
    }
  for (i=1; i<=n; i++) {
    key3 = component[i];
    if (key3 < 0) key3 = i;
    component[i] = aux[key3];
  }
  if (numcomp == 1) {
    component[0] = numcomp;
    return;
  }
  typeb = numnodes[1];
```

```
numnodes[1] = 1;
for (i=2; i<=numcomp; i++) {
  typea = numnodes[i];
  numnodes[i] = numnodes[i-1] + typeb - 1;
  typeb = typea;
}
for (i=1; i<=edges; i++) {
  typec = nodei[i];
  compkey = component[typec];
  aux[i] = numnodes[compkey];
  numnodes[compkey]++;
}
for (i=1; i<=q; i++) {
  typea = index[i];
  do {
    if (typea <= i) break;
    typeb = index[typea];
    index[typea] = -typeb;
    typea = typeb;
  } while (true);
  index[i] = -index[i];
}
if (aux[1] >= 0)
  for (j=1; j<=edges; j++) {
    tracka = aux[j];
    do {
      if (tracka <= j) break;
      trackb = aux[tracka];
      aux[tracka] = -trackb;
      tracka = trackb;
    } while (true);
    aux[j] = -aux[j];
  }
for (i=1; i<=q; i++) {
  typea = -index[i];
  if(typea >= 0) {
    r = 0;
    do {
      typea = index[typea];
      r++;
    } while (typea > 0);
    typea = i;
    for (j=1; j<=edges; j++)
      if (aux[j] <= 0) {
        trackb = j;
        p = r;
        do {
          tracka = trackb;
          key1 = (typea == 1) ? nodei[tracka] : nodej[tracka];
          do {
```

```
                typea = Math.abs(index[typea]);
                key1 = (typea == 1) ? nodei[tracka] : nodej[tracka];
                tracka = Math.abs(aux[tracka]);
                key2 = (typea == 1) ? nodei[tracka] : nodej[tracka];
                if (typea == 1)
                  nodei[tracka] = key1;
                else
                  nodej[tracka] = key1;
                key1 = key2;
                if (tracka == trackb) {
                  p--;
                  if (typea == i) break;
                }
              } while (true);
              trackb = Math.abs(aux[trackb]);
            } while (p != 0);
          }
      }
    }
    for (i=1; i<=q; i++)
      index[i] = Math.abs(index[i]);
    if (aux[1] > 0) {
      component[0] = numcomp;
      return;
    }
    for (j=1; j<=edges; j++)
      aux[j] = Math.abs(aux[j]);
    typea=1;
    for (i=1; i<=numcomp; i++) {
      typeb = numnodes[i];
      numnodes[i] = typeb - typea + 1;
      typea = typeb;
    }
    component[0] = numcomp;
}
```

Example:

Identify the different components of the following graph.

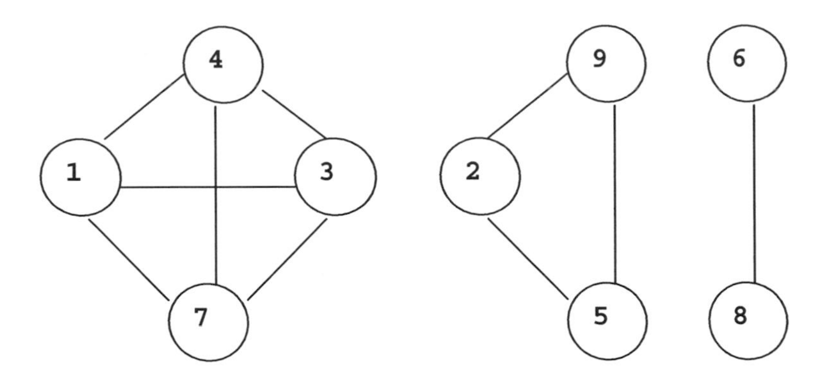


```
package GraphAlgorithms;

public class Test_connectedComponents extends Object {

  public static void main(String args[]) {

    int n=9, m=10;
    int component[] = new int[n+1];
    int nodei[] = {0,4,9,1,5,4,3,9,7,6,1};
    int nodej[] = {0,3,5,7,2,1,7,2,4,8,3};

    GraphAlgo.connectedComponents(n,m,nodei,nodej,component);
    System.out.println("Total number of components = " + component[0]);
    System.out.print("\n              Node: ");
    for (int i=1; i<=n; i++)
      System.out.print("  " + i);
    System.out.print("\n Component Number: ");
    for (int i=1; i<=n; i++)
      System.out.print("  " + component[i]);
    System.out.println();
  }
}
```

Output:

```
Total number of components = 3

              Node:  1  2  3  4  5  6  7  8  9
 Component Number:   1  3  1  1  3  2  1  2  3
```

2.6 Cut Nodes

An undirected graph that is not connected can be divided into connected *components*. A connected graph consists of one single component. A *cut node* of a component is a node whose removal will disconnect the component. The following procedure [P71] finds the number of components and all the cut nodes for a given simple undirected graph of n nodes in time $O(n^2)$. The set of nodes V of the input graph G are assumed to be numbered from 1 to n. A tree T rooted at a node r will be grown to span G.

Let p(i) denote the unique predecessor of each node i in the tree, d(i) be the distance from node i to the root of T, b(i) be the label assigned to the edge (i,p(i)), and h(j) be a Boolean variable for marking edge j. For each component of the given graph, perform the following:

Step 1. Set b(i) = 0, h(i) = false, for all i. Choose an arbitrary node r as the root. Initially T consists of the single node r,
d(r) = 0, X is empty, Y = T, Z = V – {r}.

Step 2. If Y is nonempty then select the most recent member of Y, say u, delete u from Y and continue with Step 3. If Y is empty then count the number of blocks of G by noting that edge (i,p(i)) belongs to block b(i). If b(i) is equal to –i then the edge itself is a block of G. If G has only one block then stop, otherwise G has at least one cut node. The cut nodes are identified by the property that node i is a cut node if and only if there are two or more distinct labels on edges of T through i. Stop.

Step 3. Set L = 0. For each edge (u,v), where v ∈ Z, do the following:
add(u,v) to T and move v from Z to Y,
set p(v) = u, d(v) = d(u) + 1, b(v) = –v.
For each edge (u,v), where v ∈ Y, do the following:
if b(v) > 0 then set h(b(v)) = true,
set b(v) = u and L = max(L, d(u) – d(p(v))).

Step 4. If L = 0 then add u to X and return to Step 2. Otherwise, for each edge, say (i,p(i)), on the path of length L from u to the root of the tree, set
h(b(i)) = true and b(i) = u.
For each edge (j,p(j)) for which h(b(j)) = true, set b(j) = u. Add u to X and return to Step 2.

Procedure parameters:

int cutNodes (*n, m, nodei, nodej, cutnode*)

cutNodes: int;
exit: returns the number of components of the graph.

n: int;
entry: number of nodes of the simple undirected graph, nodes of the graph are labeled from 1 to *n*.

m: int;
entry: number of edges of the simple undirected graph.

nodei, nodej: int[*m*+1];
entry: *nodei[p]* and *nodej[p]* are the end nodes of the p-th edge in the graph, p=1,2,...,*m*. The graph is not necessarily connected.

cutnode: int[*n*+1];
exit: *cutnode[0]* gives the number of cut nodes of the graph, and the cut nodes are given by *cutnode[i]*, for i=1,2,..., *cutnode[0]*.

```
public static int cutNodes(int n, int m, int nodei[], int nodej[], int cutnode[])
{
  int i,j,k,nodeu,nodev,node1,node2,node3,node4,numblocks;
  int root,p,edges,index,len1,len2,low,up,components;
  int totalcutnodes,numcutnodes=0;
  int firstedges[] = new int[n+1];
  int label[] = new int[n+1];
  int nextnode[] = new int[n+1];
  int length[] = new int[n+1];
  int cutvertex[] = new int[n+1];
  int cutedge[] = new int[m+1];
  boolean mark[] = new boolean[n+1];
  boolean join,iterate;

  totalcutnodes = 0;
  for (i=1; i<=n; i++)
    nextnode[i] = 0;
  components = 0;
  for (root=1; root<=n; root++) {
    if (nextnode[root] == 0) {
      components++;
      // set up the forward star representation of the graph
      k = 0;
      for (i=1; i<=n-1; i++) {
        firstedges[i] = k + 1;
        for (j=1; j<=m; j++) {
          nodeu = nodei[j];
          nodev = nodej[j];
          if ((nodeu == i) && (nodeu < nodev)) {
            k++;
            cutedge[k] = nodev;
          }
          else {
            if ((nodev == i) && (nodev < nodeu)) {
              k++;
              cutedge[k] = nodeu;
            }
          }
        }
      }
      firstedges[n] = m + 1;
      for (i=1; i<=n; i++) {
        label[i] = 0;
```

```
      mark[i] = false;
    }
    length[root] = 0;
    nextnode[root] = -1;
    label[root] = -root;
    index = 1;
    cutvertex[1] = root;
    edges = 2;
    do {
      node3 = cutvertex[index];
      index--;
      nextnode[node3] = -nextnode[node3];
      len1 = 0;
      for (node2=1; node2<=n; node2++) {
        join = false;
        if (node2 != node3) {
          if (node2 < node3) {
            nodeu = node2;
            nodev = node3;
          }
          else {
            nodeu = node3;
            nodev = node2;
          }
          low = firstedges[nodeu];
          up = firstedges[nodeu + 1];
          if (up > low) {
            up--;
            for (k=low; k<=up; k++)
              if (cutedge[k] == nodev) {
                join = true;
                break;
              }
          }
        }
        if (join) {
          node1 = nextnode[node2];
          if (node1 == 0) {
            nextnode[node2] = -node3;
            index++;
            cutvertex[index] = node2;
            length[node2] = length[node3] + 1;
            label[node2] = -node2;
          }
          else {
            if (node1 < 0) {
              // next block
              node4 = label[node2];
              if (node4 > 0) mark[node4] = true;
              label[node2] = node3;
```

```
          len2 = length[node3] - length[-node1];
          if (len2 > len1)  len1 = len2;
        }
      }
    }
  }
  if (len1 > 0) {
    j = node3;
    while (true) {
      len1--;
      if (len1 < 0) break;
      p = label[j];
      if (p > 0) mark[p] = true;
      label[j] = node3;
      j = nextnode[j];
    }
    for (i=1; i<=n; i++) {
      p = label[i];
      if (p > 0)
        if (mark[p]) label[i] = node3;
    }
  }
  edges++;
} while ((edges <= n) && (index > 0));
nextnode[root] = 0;
node3 = cutvertex[1];
nextnode[node3] = Math.abs(nextnode[node3]);
numblocks = 0;
numcutnodes = 0;
for (i=1; i<=n; i++)
  if (i != root) {
    node3 = label[i];
    if (node3 < 0) {
      numblocks++;
      label[i] = n + numblocks;
    }
    else {
      if ((node3 <= n) && (node3 > 0)) {
        numblocks++;
        node4 = n + numblocks;
        for (j=i; j<=n; j++)
          if (label[j] == node3) label[j] = node4;
      }
    }
  }
for (i=1; i<=n; i++) {
  p = label[i];
  if (p > 0) label[i] = p - n;
}
i = 1;
```

```
        while (nextnode[i] != root)
          i++;
        label[root] = label[i];
        for (i=1; i<=n; i++) {
          node1 = nextnode[i];
          if (node1 > 0) {
            p = Math.abs(label[node1]);
            if (Math.abs(label[i]) != p) label[node1] = -p;
          }
        }
        for (i=1; i<=n; i++)
          if (label[i] < 0) numcutnodes++;
        // store the cut nodes
        j = 0;
        for (i=1; i<=n; i++)
          if (label[i] < 0) {
            j++;
            cutvertex[j] = i;
          }
        //  find the end-nodes
        for (i=1; i<=n; i++)
          length[i] = 0;
        for (i=1; i<=m; i++) {
          j = nodei[i];
          length[j]++;
          j = nodej[i];
          length[j]++;
        }
        for (i=1; i<=n; i++)
          if (length[i] == 1)
            if (label[i] > 0) label[i] = -label[i];
        for (p=1; p<=numcutnodes; p++) {
          totalcutnodes++;
          cutnode[totalcutnodes] = cutvertex[p];
        }
      }
    }
    cutnode[0] = totalcutnodes;
    return components;
  }
```

Example:

Find the number of components and the cut nodes of the following graph.

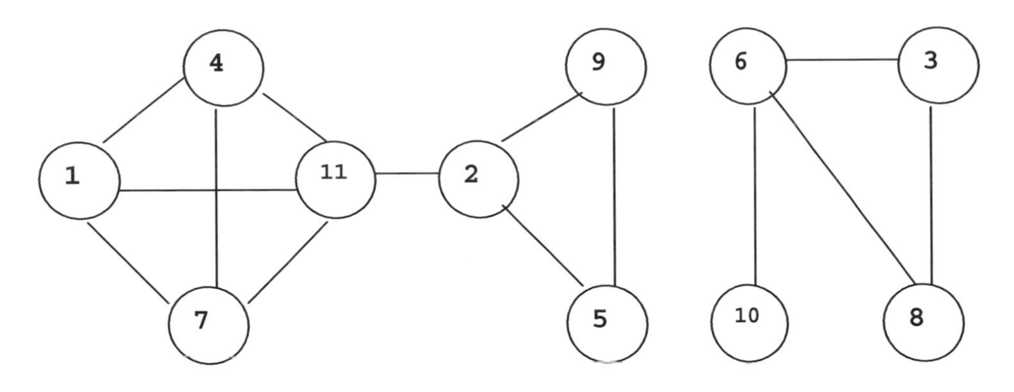


```
package GraphAlgorithms;

public class Test_cutNodes extends Object {

  public static void main(String args[]) {

    int k;
    int n=11;
    int m=14;
    int nodei[] = {0, 4,9,1,3,6,5,4, 2,11,9,7, 6, 1,6};
    int nodej[] = {0,11,5,7,8,8,2,1,11, 7,2,4,10,11,3};
    int cutnode[] = new int[n+1];

    k = GraphAlgo.cutNodes(n,m,nodei,nodej,cutnode);
    System.out.print("number of components = " + k +
                "\n  number of cutnodes = " + cutnode[0] +
                "\n\nThe cut nodes are: ");
    for (int i=1; i<=cutnode[0]; i++)
      System.out.printf("%4d", cutnode[i]);
    System.out.println();
  }
}
```

Output:

```
number of components = 2
  number of cutnodes = 3

The cut nodes are:    2  11   6
```

2.7 Strongly Connected Components

A *strongly connected component* of a directed graph is a maximal set of nodes in which there is a directed path from any one node in the set to any other node in the set. The following procedure [T72, TS92] finds the strongly connected components of a directed graph of n nodes and m edges by a depth-first search with a running time of $O(\max(n,m))$.

Step 1. Select one node s in G as a starting node and mark s as "visited". Each unvisited node adjacent to s is searched in turn using the depth-first search. The search of s is completed when all nodes that can be reached from s have been visited. If some nodes remain unvisited then an unvisited node is arbitrarily selected as a new starting node. This process is repeated until all nodes of G have been visited.

Step 2. Let r_1, r_2, ..., r_k be the roots in the order in which the depth-first search of the nodes terminated. The strongly connected component G_1 with root r_1 consists of all descendants of r_1. Furthermore, for each i, i = 2, 3, ..., k, the strongly connected component G_i with root r_i consists of the nodes which are descendants of r_i but are in none of G_1, G_2, ..., G_{i-1}.

Procedure parameters:

void stronglyConnectedComponents (*n, m, nodei, nodej, component*)

n: int;
entry: number of nodes of the directed graph,
nodes of the graph are labeled from 1 to *n*.

m: int;
entry: number of edges of the directed graph.

nodei, nodej: int[*m*+1];
entry: *nodei[p]* and *nodej[p]* are the end nodes of the p-th edge in the graph, p=1,2,...,*m*.

component: int[*n*+1];
exit: *component[0]* gives the number of strongly connected components of the graph.
component[i] is the number of the component to which node i belongs, for i=1,2,...,*n*.

```
public static void stronglyConnectedComponents(int n, int m, int nodei[],
                                              int nodej[], int component[])
{
  int i,j,k,series,stackpointer,numcomponents,p,q,r;
  int backedge[] = new int[n+1];
  int parent[] = new int[n+1];
  int sequence[] = new int[n+1];
  int stack[] = new int[n+1];
  int firstedges[] = new int[n+2];
  int endnode[] = new int[m+1];
  boolean next[] = new boolean[n+1];
```

```
boolean trace[] = new boolean[n+1];
boolean fresh[] = new boolean[m+1];
boolean skip,found;

// set up the forward star representation of the graph
firstedges[1] = 0;
k = 0;
for (i=1; i<=n; i++) {
  for (j=1; j<=m; j++)
    if (nodei[j] == i) {
      k++;
      endnode[k] = nodej[j];
    }
  firstedges[i+1] = k;
}
for (j=1; j<=m; j++)
  fresh[j] = true;
// initialize
for (i=1; i<=n; i++) {
  component[i] = 0;
  parent[i] = 0;
  sequence[i] = 0;
  backedge[i] = 0;
  next[i] = false;
  trace[i] = false;
}
series = 0;
stackpointer = 0;
numcomponents = 0;
// choose an unprocessed node not in the stack
while (true) {
  p = 0;
  while (true) {
    p++;
    if (n < p) {
      component[0] = numcomponents;
      return;
    }
    if (!trace[p]) break;
  }
  series++;
  sequence[p] = series;
  backedge[p] = series;
  trace[p] = true;
  stackpointer++;
  stack[stackpointer] = p;
  next[p] = true;
  while (true) {
    skip = false;
    for (q=1; q<=n; q++) {
```

```
        // find an unprocessed edge (p,q)
        found = false;
        for (i=firstedges[p]+1; i<=firstedges[p+1]; i++)
          if ((endnode[i] == q) && fresh[i]) {
            // mark the edge as processed
            fresh[i] = false;
            found = true;
            break;
          }
        if (found) {
          if (!trace[q]) {
            series++;
            sequence[q] = series;
            backedge[q] = series;
            parent[q] = p;
            trace[q] = true;
            stackpointer++;
            stack[stackpointer] = q;
            next[q] = true;
            p = q;
          }
          else {
            if (trace[q]) {
              if (sequence[q] < sequence[p] && next[q]) {
                backedge[p] = (backedge[p] < sequence[q]) ?
                               backedge[p] : sequence[q];
              }
            }
          }
          skip = true;
          break;
        }
      }
      if (skip) continue;
      if (backedge[p] == sequence[p]) {
        numcomponents++;
        while (true) {
          r = stack[stackpointer];
          stackpointer--;
          next[r] = false;
          component[r] = numcomponents;
          if (r == p) break;
        }
      }
      if (parent[p] != 0) {
        backedge[parent[p]] = (backedge[parent[p]] < backedge[p]) ?
                               backedge[parent[p]] : backedge[p];
        p = parent[p];
      }
      else
```

```
        break;
      }
    }
  }
```

Example:

Find the strongly connected components of the following graph.

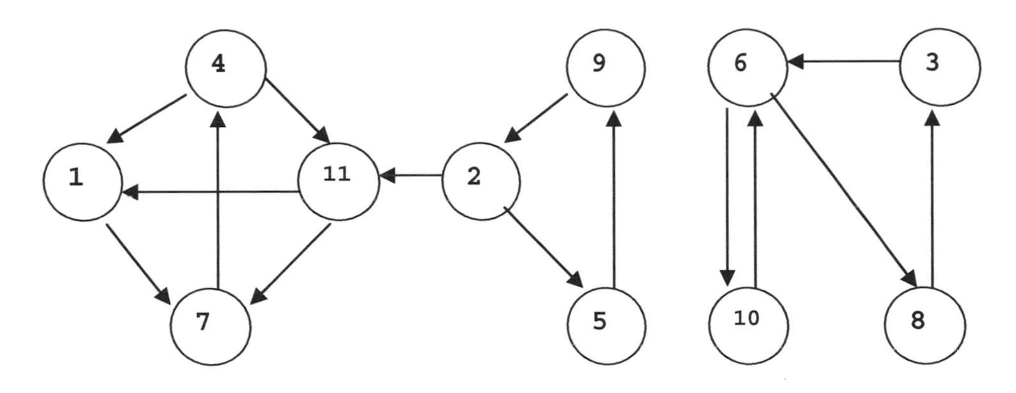


```
package GraphAlgorithms;

public class Test_stronglyConnectedComponents extends Object {

  public static void main(String args[]) {
    int n=11;
    int m=14;
    int nodei[] = {0,7,10,1,6,3, 4,2,11,8,9, 6,4, 2,5};
    int nodej[] = {0,4, 6,7,8,6,11,5, 7,3,2,10,1,11,9};
    int component[] = new int[n+1];

    GraphAlgo.stronglyConnectedComponents(n,m,nodei,nodej,component);
    System.out.println("Number of strongly connected components = " +
                       component[0]);
    for (int i=1; i<=component[0]; i++) {
      System.out.printf("\n Nodes in component " + i + ": ");
      for (int j=1; j<=n; j++)
        if (component[j] == i)
          System.out.printf("%3d",j);
    }
    System.out.println();
  }
}
```

Output:

```
Number of strongly connected components = 3

 Nodes in component 1:   1  4  7 11
 Nodes in component 2:   2  5  9
 Nodes in component 3:   3  6  8 10
```

2.8 Minimal Equivalent Graph

The *minimal equivalent graph* problem is to find a directed subgraph H from a given strongly connected graph G by removing the maximum number of edges from G without affecting its reachability properties. That is, for any two nodes u and v of G, there is a directed path from u to v in H. Let E be the set of edges of a given strongly connected graph G of n nodes and m edges. The following branch and bound algorithm [M77] finds a minimal equivalent graph of G with running time $O(2^m)$.

Step 1. Examine all edges sequentially. An edge (i,j) is removed from G whenever there exists an alternative path from node i to node j which does not include previously eliminated edges. Let F be the set of edges which are removed from G, and S = E – F be the current solution.

Step 2. Iteratively execute a combination of backtracking and forward moves. A backtracking move will remove from F the highest labeled edge, say the k-th one, and add it back in E. This backtracking move is followed by a forward move which will sequentially consider all edges in E with labels greater than k, removing from E every edge (i,j) for which an alternative path exists, and adding (i,j) back to F.

Step 3. If $|E| < |S|$, then update the current solution by setting S = E, and return to Step 2. The algorithm terminates when no further backtracking is possible in Step 2 (in which case F is empty) or when $|S| = n$.

Procedure parameters:

void minimalEquivalentGraph (*n, m, nodei, nodej, link*)

n: int;
entry: number of nodes of the directed graph,
nodes of the graph are labeled from 1 to *n*.

m: int;
entry: number of edges of the directed graph.

nodei, nodej: int[*m*+1];
entry: *nodei[p]* and *nodej[p]* are the end nodes of the p-th edge in the strongly connected graph, p=1,2,...,*m*.

link: boolean[*m*+1];
exit: *link[i]* = true if the i-th edge is in the minimal equivalent graph, otherwise *link[i]* = false, for i=1,2,..., *m*.

```
public static void minimalEquivalentGraph(int n, int m, int nodei[],
                                          int nodej[], boolean link[])
{
  int i,j,k,nodeu,nodev,n1,low,up,edges,index1,index2,high,kedge=0;
  int nextnode[] = new int[n+1];
  int ancestor[] = new int[n+1];
  int descendant[] = new int[n+1];
  int firstedges[] = new int[n+2];
  int pointer[] = new int[m+1];
  int endnode[] = new int[m+1];
  boolean pathexist[] = new boolean[n+1];
  boolean currentarc[] = new boolean[m+1];
  boolean pexist[] = new boolean[1];
  boolean join,skip,hop;

  n1 = n + 1;
  // set up the forward star representation of the graph
  k = 0;
  for (i=1; i<=n; i++) {
    firstedges[i] = k + 1;
    for (j=1; j<=m; j++)
      if (nodei[j] == i) {
        k++;
        pointer[k] = j;
        endnode[k] = nodej[j];
        currentarc[k] = true;
      }
  }
  firstedges[n1] = m + 1;
  // compute number of descendants and ancestors of each node
  for (i=1; i<=n; i++) {
    descendant[i] = 0;
    ancestor[i] = 0;
  }
  edges = 0;
  for (k=1; k<=m; k++) {
    i = nodei[k];
    j = nodej[k];
    descendant[i]++;
    ancestor[j]++;
    edges++;
  }
  if (edges == n) {
    for (k=1; k<=m; k++)
      link[pointer[k]] = currentarc[k];
    return;
  }
  index1 = 0;
  for (k=1; k<=m; k++) {
    i = nodei[pointer[k]];
```

```
    j = nodej[pointer[k]];
    // check for the existence of an alternative path
    if (descendant[i] != 1) {
      if (ancestor[j] != 1) {
        currentarc[k] = false;
        minimalEqGraphFindp(n,m,n1,i,j,endnode,firstedges,
                     currentarc,pexist,nextnode,pathexist);
        if (pexist[0]) {
          descendant[i]--;
          ancestor[j]--;
          index1++;
        }
        else
          currentarc[k] = true;
      }
    }
  }
  if (index1 == 0) {
    for (k=1; k<=m; k++)
      link[pointer[k]] = currentarc[k];
    return;
  }
  high = 0;
  nodeu = n;
  nodev = n;
  // store the current best solution
  iterate:
  while (true) {
    for (k=1; k<=m; k++)
      link[k] = currentarc[k];
    index2 = index1;
    if ((edges - index2) == n) {
      for (k=1; k<=m; k++)
        currentarc[k] = link[k];
      for (k=1; k<=m; k++)
        link[pointer[k]] = currentarc[k];
      return;
    }
    // forward move
    while (true) {
      join = false;
      low = firstedges[nodeu];
      up = firstedges[nodeu + 1];
      if (up > low) {
        up--;
        for (k=low; k<=up; k++)
          if (endnode[k] == nodev) {
            join = true;
            kedge = k;
            break;
```

```
    }
}
hop = false;
if (join) {
  if (!currentarc[kedge]) {
    currentarc[kedge] = true;
    descendant[nodeu]++;
    ancestor[nodev]++;
    index1--;
    if (index1 + high - (n - nodeu) > index2) {
      if (nodev != n)
        nodev++;
      else {
        if (nodeu != n) {
          nodeu++;
          nodev = 1;
        }
        else
          continue iterate;
      }
      while (true) {
        // backtrack move
        join = false;
        low = firstedges[nodeu];
        up = firstedges[nodeu + 1];
        if (up > low) {
           up--;
           for (k=low; k<=up; k++)
             if (endnode[k] == nodev) {
               join = true;
               kedge = k;
               break;
             }
        }
        if (join) {
          high--;
          skip = false;
          if (descendant[nodeu] != 1) {
            if (ancestor[nodev] != 1) {
              currentarc[kedge] = false;
              minimalEqGraphFindp(n,m,n1,nodeu,nodev,endnode,
                firstedges,currentarc,pexist,nextnode,pathexist);
              if (pexist[0]) {
                descendant[nodeu]--;
                ancestor[nodev]--;
                index1++;
                skip = true;
              }
              else
                currentarc[kedge] = true;
```

```
                }
              }
              if (!skip) {
                if (index1 + high - (n - nodeu) <= index2) {
                  high++;
                  hop = true;
                  break;
                }
              }
              // check for the termination of the forward move
              if (high - (n - nodeu) == 0) continue iterate;
            }
            if (nodev != n)
              nodev++;
            else {
              if (nodeu != n) {
                nodeu++;
                nodev = 1;
              }
              else
                continue iterate;
            }
          }
        }
      }
      if (!hop) high++;
    }
    hop = false;
    if (nodev != 1) {
      nodev--;
      continue;
    }
    if (nodeu == 1) {
      for (k=1; k<=m; k++)
        currentarc[k] = link[k];
      for (k=1; k<=m; k++)
        link[pointer[k]] = currentarc[k];
      return;
    }
    nodeu--;
    nodev = n;
  }
 }
}


static private void minimalEqGraphFindp(int n, int m, int n1, int nodeu,
       int nodev, int endnode[], int firstedges[], boolean currentarc[],
       boolean pexist[], int nextnode[], boolean pathexist[])
{
```

```
/* this method is used internally by minimalEquivalentGraph */

// determine if a path exists from nodeu to nodev by Yen's algorithm

int i,j,k,i2,j2,low,up,kedge=0,index1,index2,index3;
boolean join;

// initialization
for (i=1; i<=n; i++) {
  nextnode[i] = i;
  pathexist[i] = false;
}
pathexist[nodeu] = true;
nextnode[nodeu] = n;
index1 = nodeu;
index2 = n - 1;
// compute the shortest distance labels
i = 1;
while (true) {
  j = nextnode[i];
  join = false;
  low = firstedges[index1];
  up = firstedges[index1 + 1];
  if (up > low) {
    up--;
    for (k=low; k<=up; k++)
      if (endnode[k] == j) {
        join = true;
        kedge = k;
        break;
      }
  }
  if (join)
    if (currentarc[kedge]) pathexist[j] = true;
  if (pathexist[j]) {
    index3 = i + 1;
    if (index3 <= index2) {
      for (i2=index3; i2<=index2; i2++) {
        j2 = nextnode[i2];
        join = false;
        low = firstedges[index1];
        up = firstedges[index1 + 1];
        if (up > low) {
          up--;
          for (k=low; k<=up; k++)
            if (endnode[k] == j2) {
              join = true;
              kedge = k;
              break;
            }
```

```
          }
          if (join)
            if (currentarc[kedge]) pathexist[j2] = true;
        }
      }
      // check whether an alternative path exists
      if (pathexist[nodev]) {
         pexist[0] = true;
         return;
      }
      nextnode[i] = nextnode[index2];
      index1 = j;
      index2--;
      if (index2 > 1) continue;
      join = false;
      low = firstedges[index1];
      up = firstedges[index1 + 1];
      if (up > low) {
        up--;
        for (k=low; k<=up; k++)
          if (endnode[k] == nodev) {
            join = true;
            kedge = k;
            break;
          }
      }
      pexist[0] = false;
      if (join)
        if (currentarc[kedge]) pexist[0] = true;
      return;
    }
    i++;
    if (i <= index2) continue;
    pexist[0] = false;
    return;
  }
}
```

Example:

Find a minimal equivalent graph of the following strongly connected digraph.

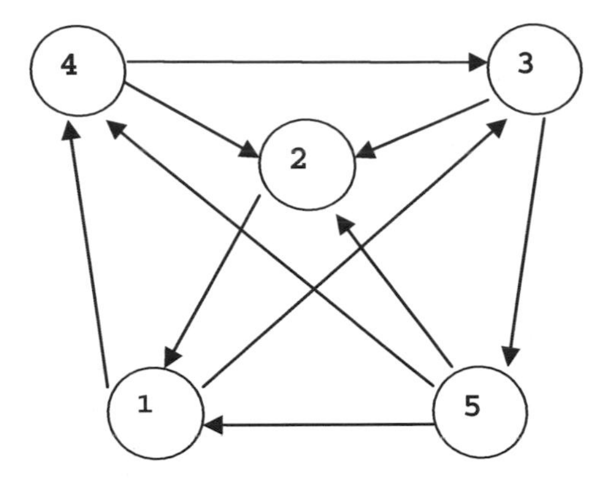


```
package GraphAlgorithms;

public class Test_minimalEquivalentGraph extends Object {

  public static void main(String args[]) {
    int n=5;
    int m=10;
    int nodei[] = {0,4,5,3,1,5,1,4,2,3,5};
    int nodej[] = {0,3,1,2,4,2,3,2,1,5,4};
    boolean arc[] = new boolean[m+1];

    GraphAlgo.minimalEquivalentGraph(n,m,nodei,nodej,arc);
    System.out.print("Edges of the minimal equivalent graph:\n");
    for (int k=1; k<=m; k++)
      if (arc[k])
        System.out.printf(" %4d%3d\n", nodei[k], nodej[k]);
  }
}
```

Output:

```
Edges of the minimal equivalent graph:
    4  3
    5  1
    1  4
    5  2
    2  1
    3  5
```

2.9 Edge Connectivity

The *edge connectivity* of an undirected graph is the minimum number of edges whose removal will result in a disconnected graph or a graph with a single node. The following procedure [ET75] finds the edge connectivity of a connected undirected graph of n nodes by solving n maximum flow network problems. Since the maximum network flow problem requires $O(n^3)$ operations [K74], the edge connectivity will be found in $O(n^4)$ operations.

Denote the nodes of the given connected undirected graph G by 1, 2, ..., n.
For j=2 to n do the following:

Take node 1 as the source, node j as the sink in G, assign a unit capacity to all edges in both directions, and find the value of a maximum flow f(j) in the resulting network. The edge connectivity is equal to the minimum of all f(j), for j=2,3,...,n.

Procedure parameters:

int edgeConnectivity (*n, m, nodei, nodej*)

edgeConnectivity: int;
exit: the edge connectivity of the graph.
n: int;
entry: the number of nodes of the undirected graph, labeled from 1 to *n*.
m: int;
entry: the number of edges of the graph.
nodei, nodej: int[*m*+1];
entry: *nodei[i]* and *nodej[i]* are the end nodes of the i-th edge in the graph, i=1,2,...,*m*.

```
public static int edgeConnectivity(int n, int m, int nodei[], int nodej[])
{
  int i,j,k,m2,source,sink;
  int minimumcut[] = new int[n+1];
  int edgei[] = new int[4*m+1];
  int edgej[] = new int[4*m+1];
  int capac[] = new int[4*m+1];
  int arcflow[] = new int[4*m+1];
  int nodeflow[] = new int[4*m+1];

  k = n;
  source = 1;
  m2 = m + m;
  for (sink=2; sink<=n; sink++) {
    // construct the network
    for (i=1; i<=4*m; i++) {
      edgei[i] = 0;
      edgej[i] = 0;
```

```
        capac[i] = 0;
      }
      // duplicate the edges
      j = 0;
      for (i=1; i<=m; i++) {
        j++;
        edgei[j] = nodei[i];
        edgej[j] = nodej[i];
        capac[j] = 1;
        j++;
        edgei[j] = nodej[i];
        edgej[j] = nodei[i];
        capac[j] = 1;
      }
      // invoke the network flow algorithm
      maximumNetworkFlow(n,m2,edgei,edgej,capac,source,sink,
                         minimumcut,arcflow,nodeflow);
      if (nodeflow[source] < k) k = nodeflow[source];
    }
    return k;
  }
```

Example:

Find the edge connectivity of the following graph.

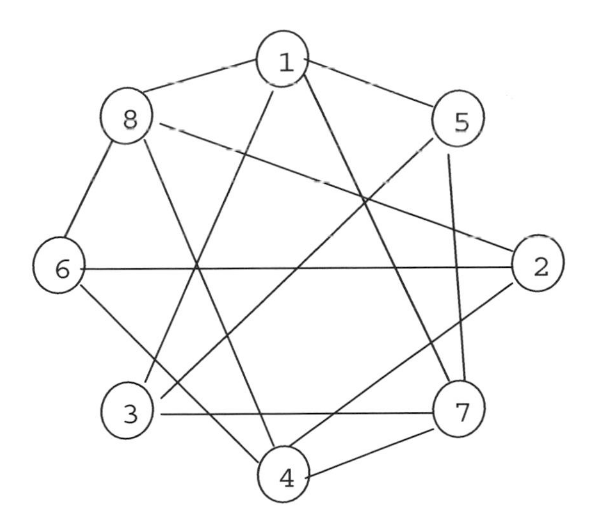


```
package GraphAlgorithms;

public class Test_edgeConnectivity extends Object {

  public static void main(String args[]) {

    int k;
```

```
      int n = 8;
      int m = 13;
      int nodei[] = {0,4,8,3,1,2,4,3,3,4,6,5,8,1};
      int nodej[] = {0,7,1,7,5,8,6,7,1,2,8,3,4,7};

      k = GraphAlgo.edgeConnectivity(n,m,nodei,nodej);
      System.out.println("The edge connectivity of the graph = " + k);
   }
}
```

Output:

```
The edge connectivity of the graph = 2
```

2.10 Minimum Spanning Tree

The *minimum spanning tree* problem is to find a spanning tree in a given undirected weighted graph G such that the sum of the edge weights in the tree is minimum. The following are two methods in solving this problem.

Prim's Method

Chooses any node j and let this single node j be the partially constructed tree T. Join to T an edge of G whose weight is minimal among all edges with one end node in T and the other end node not in T. Repeat this process until T becomes a spanning tree of G. This algorithm is due to Prim [P57], and has the running time of $O(n^2)$ for a given graph of n nodes.

Kruskal's Method

Initialize an empty set T. Edges are considered to be included in T in the nondecreasing order of the edge weights. An edge is included in T if it does not form a cycle with the edges already in T. A minimum spanning tree is formed when $n - 1$ edges are included in T, where n is the number of nodes in G.

In the implementation of this method, the edges are partially sorted with the smallest weight edge at the root of a heap structure (a binary tree in which the weight of every node is not greater than the weights of its descendants). This algorithm is due to Kruskal [K56], and has the running time of $O(m \log m)$ for a given graph of m edges.

Procedure parameters (Prim's method):

void minimumSpanningTreePrim (*n, dist, tree*)

n: int;
entry: number of nodes of the simple undirected graph, nodes of the graph are labeled from 1 to *n*.

dist: int[*n*+1][*n*+1];
entry: *dist[i][j]* is the weight of the edge between node i and node j, for i=1,2,...,*n* and j=1,2,...,*n*. Note that *dist[i][i]* = 0.

tree: int[*n*+1];
exit: The two end nodes of the i-th edge of the minimum spanning tree are node "i" and node "tree[i]", for i=1,2,...,*n*–1.

```
public static void minimumSpanningTreePrim(int n, int dist[][], int tree[])
{
  int i,j,n1,d,mindist,node,k=0;

  n1 = n - 1;
  for (i=1; i<=n1; i++)
    tree[i] = -n;
  tree[n] = 0;
  for (i=1; i<=n1; i++) {
    mindist = Integer.MAX_VALUE;
    for (j=1; j<=n1; j++) {
      node = tree[j];
      if (node <= 0) {
        d = dist[-node][j];
        if (d < mindist) {
          mindist = d;
          k = j;
        }
      }
    }
    tree[k] = -tree[k];
    for (j=1; j<=n1; j++) {
      node = tree[j];
      if (node <= 0)
        if (dist[j][k] < dist[j][-node]) tree[j] = -k;
    }
  }
}
```

Example:

Find a minimum spanning tree of the following graph using Prim's method.

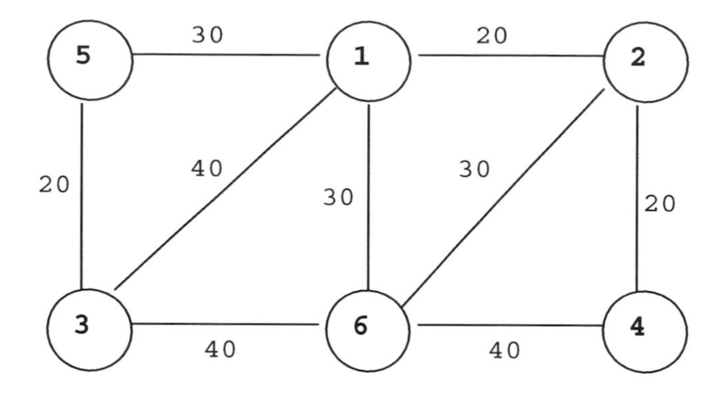


```
package GraphAlgorithms;

public class Test_minimumSpanningTreePrim extends Object {

  public static void main(String args[]) {

    int n=6;
    int tree[] = new int[n+1];
    int dist[][] = {{0,  0,  0,  0,  0,  0,  0},
                    {0,  0, 20, 99, 99, 30, 30},
                    {0, 20,  0, 99, 20, 99, 30},
                    {0, 40, 99,  0, 99, 20, 40},
                    {0, 99, 20, 99,  0, 99, 40},
                    {0, 30, 99, 20, 99,  0, 99},
                    {0, 30, 30, 40, 40, 99,  0}};

    GraphAlgo.minimumSpanningTreePrim(n, dist, tree);
    System.out.println("Minimum spanning tree edges: ");
    for (int i=1; i<=n-1; i++)
      System.out.println("  " + i + "  " + tree[i]);
  }
}
```

Output:

```
Minimum spanning tree edges:
  1 6
  2 1
  3 5
  4 2
  5 1
```

Procedure parameters (Kruskal's method):

void minimumSpanningTreeKruskal (*n,m,nodei,nodej,weight,treearc1,treearc2*)

n: int;
entry: number of nodes of the simple undirected graph, nodes of the graph are labeled from 1 to *n*.

m: int;
entry: number of edges of the simple undirected graph.

nodei, nodej: int[*m*+1];
entry: *nodei[p]* and *nodej[p]* are the end nodes of the p-th edge in the graph, p=1,2,...,*m*. The graph is not necessarily connected.

weight: int[*m*+1];
entry: *weight[i]* is the weight of the i-th edge, for i=1,2,..., *m*.

treearc1, treearc2: int[*n*];
exit: The two end nodes of the i-th edge of the minimum spanning tree are node "*treearc1[i]*" and node "*treearc2[i]*", for i=1,2,...,*n*–1.

```
public static void minimumSpanningTreeKruskal(int n, int m, int nodei[],
                int nodej[], int weight[], int treearc1[], int treearc2[])
{
  int i,index,index1,index2,index3,nodeu,nodev,nodew,len,nedge,treearc;
  int halfm,numarc,nedge2=0;
  int predecessor[] = new int[n+1];

  for (i=1; i<=n; i++)
    predecessor[i] = -1;
  // initialize the heap structure
  i = m / 2;
  while (i > 0) {
    index1 = i;
    halfm = m / 2;
    while (index1 <= halfm) {
      index = index1 + index1;
      index2 = ((index < m) && (weight[index + 1] < weight[index])) ?
                index + 1 : index;
      if (weight[index2] < weight[index1]) {
        nodeu = nodei[index1];
        nodev = nodej[index1];
        len   = weight[index1];
        nodei[index1] = nodei[index2];
        nodej[index1] = nodej[index2];
        weight[index1] = weight[index2];
        nodei[index2] = nodeu;
        nodej[index2] = nodev;
        weight[index2] = len;
        index1 = index2;
      }
      else
```

```
        index1 = m;
    }
    i--;
  }
  nedge = m;
  treearc = 0;
  numarc = 0;
  while ((treearc < n-1) && (numarc < m)) {
    // examine the next edge
    numarc++;
    nodeu = nodei[1];
    nodev = nodej[1];
    nodew = nodeu;
    // check if nodeu and nodev are in the same component
    while (predecessor[nodew] > 0) {
      nodew = predecessor[nodew];
    }
    index1 = nodew;
    nodew = nodev;
    while (predecessor[nodew] > 0) {
      nodew = predecessor[nodew];
    }
    index2 = nodew;
    if (index1 != index2) {
      // include nodeu and nodev in the minimum spanning tree
      index3 = predecessor[index1] + predecessor[index2];
      if (predecessor[index1] > predecessor[index2]) {
        predecessor[index1] = index2;
        predecessor[index2] = index3;
      }
      else {
        predecessor[index2] = index1;
        predecessor[index1] = index3;
      }
      treearc++;
      treearc1[treearc] = nodeu;
      treearc2[treearc] = nodev;
    }
    // restore the heap structure
    nodei[1] = nodei[nedge];
    nodej[1] = nodej[nedge];
    weight[1] = weight[nedge];
    nedge--;
    index1 = 1;
    nedge2 = nedge / 2;
    while (index1 <= nedge2) {
      index = index1 + index1;
      index2 = ((index < nedge) && (weight[index + 1] < weight[index])) ?
                 index + 1 : index;
      if (weight[index2] < weight[index1]) {
```

```
        nodeu = nodei[index1];
        nodev = nodej[index1];
        len   = weight[index1];
        nodei[index1] = nodei[index2];
        nodej[index1] = nodej[index2];
        weight[index1] = weight[index2];
        nodei[index2] = nodeu;
        nodej[index2] = nodev;
        weight[index2] = len;
        index1 = index2;
      }
      else
        index1 = nedge;
    }
  }
  treearc1[0] = treearc;
}
```

Example:

Find a minimum spanning tree of the following graph using Kruskal's method.

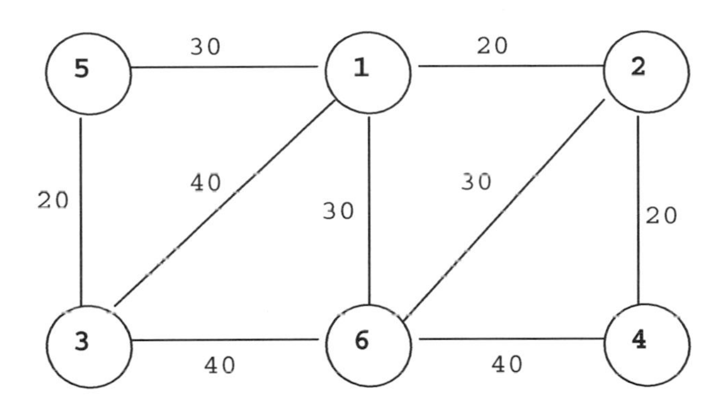


```
package GraphAlgorithms;

public class Test_minimumSpanningTreeKruskal extends Object {

  public static void main(String args[]) {

    int n=6, m=9;
    int treearc1[] = new int[n];
    int treearc2[] = new int[n];
    int nodei[] =  {0, 4, 6, 5, 1, 3, 1, 2, 1, 6};
    int nodej[] =  {0, 2, 1, 3, 2, 6, 5, 6, 3, 4};
    int weight[] = {0,20,30,20,20,40,30,30,40,40};
```

```
    GraphAlgo.minimumSpanningTreeKruskal(n,m,nodei,nodej,weight,
                                         treearc1,treearc2);
    System.out.println("Minimum spanning tree edges: ");
    for (int i=1; i<=treearc1[0]; i++)
      System.out.println("  " + treearc1[i] + " " + treearc2[i]);
  }
}
```

Output:

```
Minimum spanning tree edges:
  4 2
  1 2
  5 3
  2 6
  1 5
```

2.11 All Cliques

A *clique* is a complete subgraph of a given undirected graph G. A *k*-clique is a clique of order k. An *independent set* is a subset of nodes of G such that no two nodes of the set are adjacent in G. An independent set is maximal if there is no other independent set that contains it. Since a subset of nodes S of a graph G is a maximal independent set if and only if S is a clique in the complement of G, any algorithm which finds the maximal independent sets of a graph can also be used to find its cliques, and vice versa. The following procedure [BK73] finds all maximal independent sets by an enumerative tree search. Let P(j) be an independent set at stage j, and Q(j) be the largest set of nodes such that any node from Q(j) added to P(j) will produce an independent set P(j+1). The set Q(j) can be partitioned into two disjoint sets S(j) and T(j), where S(j) is the set of all nodes which have been used in the search to augment P(j), and T(j) is the set of all nodes which have not been used. Let E(u) be the set of nodes adjacent to u in G. The method can be described as follows.

Step 1. Set P(0) = empty, S(0) = empty, T(0) = set of nodes of G, and j = 0.

Step 2. Perform a forward branch in the tree search by choosing any node x(j) from T(j), and create three new sets:
P(j+1) = P(j) + {x(j)},
S(j+1) = S(j) – E(x(j)),
T(j+1) = T(j) – E(x(j)) – {x(j)}.
Set j = j + 1.

Step 3. If there exists a node y in S(j) such that E(y) is disjoint from T(j) then go to Step 5.

Step 4. If both S(j) and T(j) are empty then output the maximal independent set P(j) and go to Step 5. If S(j) is nonempty and T(j) is empty then go to Step 5, otherwise return to Step 2.

Step 5. Backtrack by setting j = j – 1. Remove node x(j) from P(j+1) to produce P(j). Remove node x(j) from T(j) and add it to S(j). If j = 0 and T(0) is empty then stop, otherwise return to Step 3.

Procedure parameters:

void allCliques (*n, m, nodei, nodej, clique*)

n: int;
entry: number of nodes of the simple undirected graph (not necessarily connected but without isolated nodes);
nodes of the graph are labeled from 1 to *n*.

m: int;
entry: number of edges of the simple undirected graph.

nodei, nodej: int[*m*+1];
entry: *nodei[p]* and *nodej[p]* are the end nodes of the p-th edge in the graph, p=1,2,...,*m*. The graph is not necessarily connected.

clique: int[*n*][*n*+1];
exit: *clique[0][0]* gives the total number of cliques of the graph. *clique[i][0]* gives the number of nodes of the i-th clique, and the nodes of the i-th clique are
clique[i][1], clique[i][2], ..., clique[i][clique[i][0]],
for i=1,2,..., *clique[0][0]*.

```
public static void allCliques(int n, int m, int nodei[], int nodej[],
                              int clique[][])
{
  int i,j,k,level,depth,num,numcliques,small,nodeu,nodev,nodew=0;
  int sum,p,up,low,index1,index2,indexv=0;
  int currentclique[] = new int[n+1];
  int aux1[] = new int[n+1];
  int aux2[] = new int[n+1];
  int notused[] = new int[n+2];
  int used[] = new int[n+2];
  int firstedges[] = new int[n+2];
  int endnode[] = new int[m+m+1];
  int stack[][] = new int[n+2][n+2];
  boolean join,skip,hop;

  // set up the forward star representation of the graph
  k = 0;
  for (i=1; i<=n; i++) {
    firstedges[i] = k + 1;
    for (j=1; j<=m; j++) {
      if (nodei[j] == i) {
        k++;
        endnode[k] = nodej[j];
      }
      if (nodej[j] == i) {
```

```
        k++;
        endnode[k] = nodei[j];
      }
    }
  }
  firstedges[n+1] = k + 1;
  level = 1;
  depth = 2;
  for (i=1; i<=n; i++)
    stack[level][i] = i;
  numcliques = 0;
  num = 0;
  used[level] = 0;
  notused[level] = n;
  while (true) {
    small = notused[level];
    nodeu = 0;
    aux1[level] = 0;
    while (true) {
      nodeu++;
      if ((nodeu > notused[level]) || (small == 0)) break;
      index1 = stack[level][nodeu];
      sum = 0;
      nodev = used[level];
      while (true) {
        nodev++;
        if ((nodev > notused[level]) || (sum >= small)) break;
        p = stack[level][nodev];
        if (p == index1)
          join = true;
        else {
          join = false;
          low = firstedges[p];
          up = firstedges[p + 1];
          if (up > low) {
            up--;
            for (k=low; k<=up; k++)
              if (endnode[k] == index1) {
                join = true;
                break;
              }
          }
        }
        // store up the potential candidate
        if (!join) {
          sum++;
          indexv = nodev;
        }
      }
      if (sum < small) {
```

```
        aux2[level] = index1;
        small = sum;
        if (nodeu <= used[level])
          nodew = indexv;
        else {
          nodew = nodeu;
          aux1[level] = 1;
        }
      }
    }
    // backtrack
    aux1[level] += small;
    while (true) {
      hop = false;
      if (aux1[level] <= 0) {
        if (level <= 1) return;
        level--;
        depth--;
        hop = true;
      }
      if (!hop) {
        index1 = stack[level][nodew];
        stack[level][nodew] = stack[level][used[level]+1];
        stack[level][used[level]+1] = index1;
        index2 = index1;
        nodeu = 0;
        used[depth] = 0;
        while (true) {
          nodeu++;
          if (nodeu > used[level]) break;
          p = stack[level][nodeu];
          if (p == index2)
             join = true;
          else {
            join = false;
            low = firstedges[p];
            up = firstedges[p + 1];
            if (up > low) {
               up--;
               for (k=low; k<=up; k++)
                 if (endnode[k] == index2) {
                   join = true;
                   break;
                 }
            }
          }
          if (join) {
            used[depth]++;
            stack[depth][used[depth]] = stack[level][nodeu];
          }
```

```
    }
    notused[depth] = used[depth];
    nodeu = used[level] + 1;
    while (true) {
      nodeu++;
      if (nodeu > notused[level]) break;
      p = stack[level][nodeu];
      if (p == index2)
        join = true;
      else {
        join = false;
        low = firstedges[p];
        up = firstedges[p + 1];
        if (up > low) {
          up--;
          for (k=low; k<=up; k++)
            if (endnode[k] == index2) {
              join = true;
              break;
            }
        }
      }
      if (join) {
        notused[depth]++;
        stack[depth][notused[depth]] = stack[level][nodeu];
      }
    }
    num++;
    currentclique[num] = index2;
    if (notused[depth] == 0) {
      // found a clique
      numcliques++;
      clique[numcliques][0] = num;
      for (i=1; i<=num; i++)
        clique[numcliques][i] = currentclique[i];
      clique[0][0] = numcliques;
    }
    else {
      if (used[depth] < notused[depth]) {
        level++;
        depth++;
        break;
      }
    }
  }
  while (true) {
    num--;
    used[level]++;
    if (aux1[level] > 1) {
      nodew = used[level];
```

```
          // look for candidate
          while (true) {
            nodew++;
            p = stack[level][nodew];
            if (p == aux2[level]) continue;
            low = firstedges[p];
            up = firstedges[p + 1];
            if (up <= low) break;
            up--;
            skip = false;
            for (k=low; k<=up; k++)
              if (endnode[k] == aux2[level]) {
                skip = true;
                break;
              }
            if (!skip) break;
          }
        }
        aux1[level]--;
        break;
      }
    }
  }
}
```

Example:

Find all the cliques of the following graph.

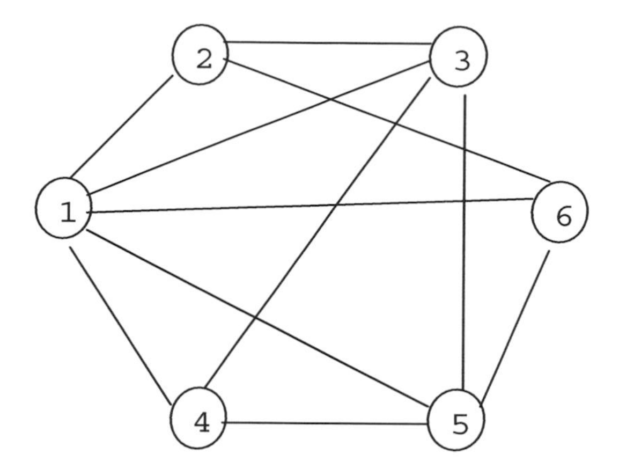


```
package GraphAlgorithms;

public class Test_allCliques extends Object {

  public static void main(String args[]) {
    int n=6;
    int m=11;
    int nodei[] = {0, 2, 4, 1, 3, 5, 3, 2, 6, 1, 2, 3};
    int nodej[] = {0, 3, 5, 6, 4, 1, 5, 1, 5, 4, 6, 1};
    int clique[][] = new int[n][n+1];

    GraphAlgo.allCliques(n,m,nodei,nodej,clique);
    System.out.println("Number of cliques = " + clique[0][0]);
    for (int i=1; i<=clique[0][0]; i++) {
      System.out.print("\nnodes of clique " + i + ": ");
      for (int j=1; j<=clique[i][0]; j++)
        System.out.printf("%3d", clique[i][j]);
    }
    System.out.println();
  }
}
```

Output:

```
Number of cliques = 4

nodes of clique 1:   1  3  4  5
nodes of clique 2:   1  3  2
nodes of clique 3:   1  6  5
nodes of clique 4:   1  6  2
```

3. Paths and Cycles

3.1 Fundamental Set of Cycles

Let T be a spanning tree of an undirected graph G with n nodes. The *fundamental set of cycles* of G corresponding to T is the set of cycles of G consisting of one edge (i,j) of G-T together with the unique path between node i and node j in T. The following procedure [K69] finds a fundamental set of cycles for a given undirected graph. Note that the graph is not necessarily connected. The method finds the fundamental set of cycles for each component H of G and has a running time of $O(n^2)$.

Step 1. Let E be the set of edges and V be the set of nodes of H. Select any node from V as the root of the tree consisting of the single node. Set T = {v}, and S = V.

Step 2. Let x be any node in $T \cap S$. If such a node does not exist then stop.

Step 3. Consider each edge (x,y) in E. If y is in T, then obtain the fundamental cycle by including the edge (x,y) together with the unique path between x and y in the tree, and delete the edge (x,y) from E. If y is not in T, then add the edge (x,y) to the tree, add node y to T, and delete the edge (x,y) from E.

Procedure parameters:

void fundamentalCycles (*n, m, nodei, nodej, fundcycle*)

n: int;
entry: number of nodes of the simple undirected graph, nodes of the graph are labeled from 1 to *n*.

m: int;
entry: number of edges of the simple undirected graph.

nodei, nodej: int[*m*+1];
entry: *nodei[p]* and *nodej[p]* are the end nodes of the p-th edge in the graph, p=1,2,...,*m*. The graph is not necessarily connected

fundcycle: int[*k*+1][*n*+1];
where *k* is the maximum number of fundamental cycles expected. If *k* is unknown, then the value of *k* can be set equal to the minimum of the two integers: m-2 and $(n-1)*(n-2)/2$.
exit: *fundcycle[0][1]* gives the number of components of the graph. *fundcycle[0][0]* gives the number of independent cycles of the graph. The number of nodes of the i-th fundamental cycle is *fundcycle[i][0]*. The nodes of the i-th fundamental cycle are
fundcycle[i][1], fundcycle[i][2], ..., fundcycle[i][fundcycle[i][0]],
for i=1,2,..., *fundcycle[0][0]*.

```
public static void fundamentalCycles(int n, int m, int nodei[], int nodej[],
                                      int fundcycle[][])
{
  int i,j,k,nodeu,nodev,components,numcycles,root,index,edges;
  int low,len,up,node1,node2,node3;
  int endnode[] = new int[m+1];
  int firstedges[] = new int[n+1];
  int nextnode[] = new int[n+1];
  int pointer[] = new int[n+1];
  int currentcycle[] = new int[n+1];
  boolean join;

  // set up the forward star representation of the graph
  k = 0;
  for (i=1; i<=n-1; i++) {
    firstedges[i] = k + 1;
    for (j=1; j<=m; j++) {
      nodeu = nodei[j];
      nodev = nodej[j];
      if ((nodeu == i) && (nodeu < nodev)) {
        k++;
        endnode[k] = nodev;
      }
      else {
        if ((nodev == i) && (nodev < nodeu)) {
          k ++;
          endnode[k] = nodeu;
        }
      }
    }
  }
  firstedges[n] = m + 1;
  for (i=1; i<=n; i++)
    nextnode[i] = 0;
  components = 0;
  numcycles = 0;
  for (root=1; root<=n; root++)
    if (nextnode[root] == 0) {
      components++;
      nextnode[root] = -1;
      index = 1;
      pointer[1] = root;
      edges = 2;
      do {
        node3 = pointer[index];
        index--;
        nextnode[node3] = -nextnode[node3];
        for (node2=1; node2<=n; node2++) {
          join = false;
          if (node2 != node3) {
```

```
    if (node2 < node3) {
      nodeu = node2;
      nodev = node3;
    }
    else {
      nodeu = node3;
      nodev = node2;
    }
    low = firstedges[nodeu];
    up = firstedges[nodeu + 1];
    if (up > low) {
      up--;
      for (k=low; k<=up; k++)
        if (endnode[k] == nodev) {
          join = true;
          break;
        }
    }
  }
  if (join) {
    node1 = nextnode[node2];
    if (node1 == 0) {
       nextnode[node2] = -node3;
       index++;
       pointer[index] = node2;
    }
    else {
      if (node1 < 0) {
        // generate the next cycle
        numcycles++;
        len = 3;
        node1 = -node1;
        currentcycle[1] = node1;
        currentcycle[2] = node2;
        currentcycle[3] = node3;
        i = node3;
        while (true) {
          j = nextnode[i];
          if (j == node1) break;
          len++;
          currentcycle[len] = j;
          i = j;
        }
        // store the current fundamental cycle
        fundcycle[numcycles][0] = len;
        for (i=1; i<=len; i++)
          fundcycle[numcycles][i] = currentcycle[i];
      }
    }
  }
```

```
        }
        edges++;
      } while ((edges <= n) && (index > 0));
      nextnode[root] = 0;
      node3 = pointer[1];
      nextnode[node3] = Math.abs(nextnode[node3]);
    }
  fundcycle[0][0] = numcycles;
  fundcycle[0][1] = components;
}
```

Example:

Find a fundamental set of cycles of the following graph.

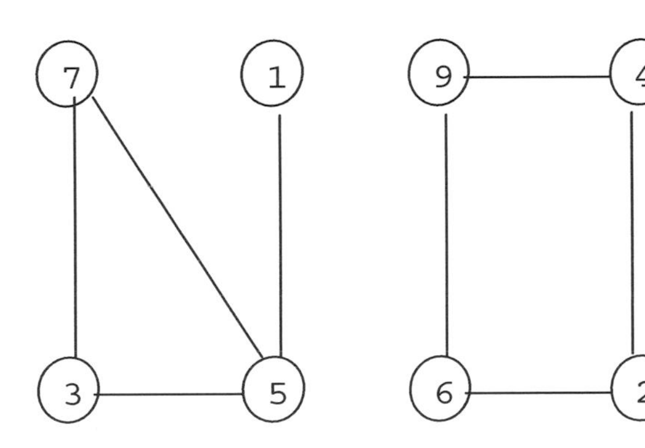


```
package GraphAlgorithms;

public class Test_fundamentalCycles extends Object {

  public static void main(String args[]) {

    int n=9;
    int m=10;
    boolean directed;
    int nodei[] = {0,4,3,6,5,8,3,4,1,9,2};
    int nodej[] = {0,2,7,2,7,4,5,9,5,6,8};
    int fundcycle[][] = new int[m-1][n+1];

    GraphAlgo.fundamentalCycles(n,m,nodei,nodej,fundcycle);
    System.out.println("number of components of the graph = " +
                       fundcycle[0][1]  + "\n");
    for (int i=1; i<=fundcycle[0][0]; i++) {
      System.out.print("nodes in cycle " + i + ": ");
      for (int j=1; j<=fundcycle[i][0]; j++)
        System.out.printf("%3d", fundcycle[i][j]);
      System.out.println();
```

```
    }
  }
}
```

Output:

```
number of components of the graph = 2

nodes in cycle 1:   5  3  7
nodes in cycle 2:   2  4  8
nodes in cycle 3:   2  4  9  6
```

3.2 Shortest Cycle Length

The *girth* is the length of the shortest cycle in a graph. The following procedure computes the girth of a given simple undirected graph of n nodes. The method starts with an adjacency matrix of the graph. Compute the square of the matrix, then the cube, etc., until a nonzero number appears in the diagonal of the matrix. The multiplication will be performed at most to the nth power.

Procedure parameters:

int girth (*n, m, nodei, nodej*)

girth: int;
exit: returns the girth of the input graph;
the return value will be zero if the graph is acyclic.

n: int;
entry: number of nodes of the simple undirected graph,
nodes of the graph are labeled from 1 to *n*.

m: int;
entry: number of edges of the simple undirected graph.

nodei, nodej: int[*m*+1];
entry: *nodei[p]* and *nodej[p]* are the end nodes of the p-th edge in the graph, p=1,2,...,*m*.

```
public static int girth(int n, int m, int nodei[], int nodej[])
{
  int diag,edges,i,j,k,nsquare,p,u,v;
  int adjlist[] = new int[n*n];
  int aux1[] = new int[n*n];
  int aux2[] = new int[n*n];
  boolean found;

  nsquare = n * n;
  // store the graph in adjlist
  for (i=0; i<nsquare; i++)
```

```
    adjlist[i] = 0;
  for (p=1; p<=m; p++) {
    i = nodei[p] - 1;
    j = nodej[p] - 1;
    adjlist[n * i + j] = 1;
  }
  // copy the adjacency list matrix
  for (i=0; i<nsquare; i++)
    aux1[i] = adjlist[i];
  // multiplication at most n times
  for (p=1; p<n; p++) {
    found = false;
    for (i=0; i<n; i++) {
      u = i * n;
      for (j=0; j<n; j++) {
        v = j * n;
        diag = u + j;
        aux2[diag] = 0;
        if (i != j) {
        for (k=0; k<n; k++)
          aux2[diag] += adjlist[u + k] * aux1[v + k];
          if (aux2[diag] > 1) {
            found = true;
            aux2[diag] = 1;
          }
        }
      }
    }
    for (i=0; i<nsquare; i++)
      if (aux1[i] + aux2[i] > 1) return 2 * p + 1;
    if (found) return 2 * (p + 1);
    // copy aux2 to aux1
    for (i=0; i<nsquare; i++)
      aux1[i] = aux2[i];
  }
  return 0;
}
```

Example:

Find the girth of the Petersen graph.

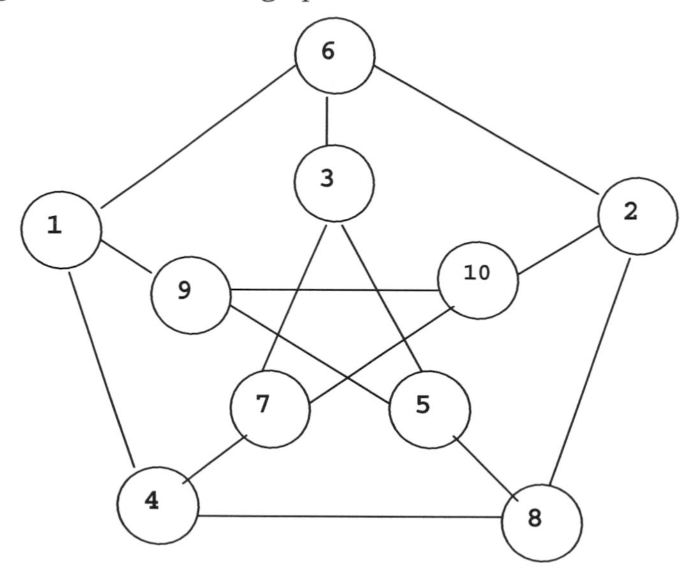


```
package GraphAlgorithms;

public class Test_girth extends Object {

  public static void main(String args[]) {

    int g;
    int n = 10;
    int m = 15;
    int nodei[] = {0,9,1,4,1,1,2,2,3,5,3,3,4, 7,10,10};
    int nodej[] = {0,5,4,7,6,9,6,8,5,8,6,7,8,10, 9,2};

    g = GraphAlgo.girth(n,m,nodei,nodej);
    System.out.println("The girth of the graph = " + g);
  }
}
```

Output:

```
The girth of the graph = 5
```

3.3 One-Pair Shortest Path

The following procedure [N66] finds a shortest path from a specified source node *s* to a specified destination node *t* in a directed graph of *n* nodes with nonnegative edge lengths. In order to reduce the computational requirements, the paths of both directions from *s* and into *t* will be considered. All paths out of *s* and into *t* of their adjacent nodes are examined simultaneously. The path which has covered the least distance is extended. The process is repeated until a path is found out of *s* which has a node on it that already existed on a path into *t*, or vice versa. The complete path is then checked to see if the shortest path is obtained. The worst-case running time of the algorithm is bounded by $O(n^2 \log n^2)$. In the thesis of DeWitt [D77], the expected running time of this algorithm on random graphs is $O(n^{1/2} \log n)$.

Procedure parameters:

int oneShortestPath (*n*, *m* , *nodei*, *nodej*, *weight*, *source*, *sink*, *path*)

oneShortestPath: int;
exit: returns the length of the shortest path from *source* to *sink* in the given directed graph.

n: int;
entry: number of nodes of the simple directed graph, nodes of the graph are labeled from 1 to *n*.

m: int;
entry: number of edges of the simple directed graph.

nodei, nodej: int[*m*+1];
entry: *nodei[p]* and *nodej[p]* are the end nodes of the p-th edge in the graph, p=1,2,...,*m*.

weight: int[*m*+1];
entry: *weight[p]* is the nonnegative length of edge p in the graph, p=1,2,...,*m*.

source: int;
entry: the source node of the graph.

sink: int;
entry: the destination node of the graph.

path: int[*n*+1];
exit: *path[0]* is the number of nodes in the shortest path from *source* to *sink* of the graph. The nodes of the shortest path are:
path[1], path[2], ..., path[path[0]].

```
public static int oneShortestPath(int n, int m, int nodei[], int nodej[],
                         int weight[], int source, int sink, int path[])
{
  int key1,key2,large,distance,index,level1,level2,lensum,temp,mnode=0;
  int i,j,num1,num2,numnodes,pathlength,minlength1,minlength2,minlength3;
  int parent[] = new int[n+1];
  int child[] = new int[n+1];
```

```
int fromsource[] = new int[n+1];
int tosink[] = new int[n+1];
int stack1[] = new int[n+1];
int stack2[] = new int[n+1];
int dist[][] = new int[n+1][n+1];

// set up the distance matrix
large = 1;
for (i=1; i<=m; i++)
  large += weight[i];
for (i=1; i<=n; i++)
  for (j=1; j<=n; j++)
    dist[i][j] = (i == j) ? 0 : large;
for (i=1; i<=m; i++)
  dist[nodei[i]][nodej[i]] = weight[i];
key1 = 0;
key2 = 0;
for (i=1; i<=n; i++) {
  fromsource[i] = dist[source][i];
  tosink[i] = dist[i][sink];
  parent[i] = source;
  child[i] = sink;
}
// find the initial values of minlength1 and minlength2 with
// corresponding 'index' values for 'fromsource' and 'tosink'
level1 = 0;
level2 = 0;
minlength2 = large;
minlength1 = large;
for (i=1; i<=n; i++) {
  // finds minlength1 and stores in stack1[1:level1] all values of k
  // such that fromsource[k] = minlength1 and fromsource[i] = key1
  temp = fromsource[i];
  if (temp > key1) {
    if (temp < minlength1) {
      level1 = 1;
      minlength1 = temp;
      stack1[level1] = i;
    }
    else {
      if (temp == minlength1) {
        level1++;
        stack1[level1] = i;
      }
    }
  }
  // finds minlength2 and stores in stack2[1:level2] all values
  // of k such that tosink[k] = minlength2 and tosink[i] = key2
  temp = tosink[i];
  if (temp > key2) {
```

```
      if (temp < minlength2) {
        level2 = 1;
        minlength2 = temp;
        stack2[level2] = i;
      }
      else {
        if (temp == minlength2) {
          level2++;
          stack2[level2] = i;
        }
      }
    }
  }
  do {
    if (minlength1 <= minlength2) {
      // reset fromsource
      key1 = minlength1;
      while (true) {
        if (level1 <= 0) break;
        index = stack1[level1];
        for (i=1; i<=n; i++) {
          distance = dist[index][i];
          lensum = minlength1 + distance;
          if (fromsource[i] > lensum) {
            fromsource[i] = lensum;
            parent[i] = index;
          }
        }
        level1--;
      }
      // find new 'minlength1' and 'index' values for 'fromsource'
      minlength1 = large;
      level1 = 0;
      for (i=1; i<=n; i++) {
        // finds minlength2 and stores in stack2[1:level2] all values
        // of k such that tosink[k] = minlength2 and tosink[i] = key2
        temp = fromsource[i];
        if (temp > key1) {
          if (temp < minlength1) {
            level1 = 1;
            minlength1 = temp;
            stack1[level1] = i;
          }
          else {
            if (temp == minlength1) {
              level1++;
              stack1[level1] = i;
            }
          }
        }
```

```
      }
    }
    else {
      // reset tosink
      key2 = minlength2;
      while (true) {
        if (level2 <= 0) break;
        index = stack2[level2];
        for (i=1; i<=n; i++) {
          distance = dist[i][index];
          lensum = minlength2 + distance;
          if (tosink[i] > lensum) {
            tosink[i] = lensum;
            child[i] = index;
          }
        }
        level2--;
      }
      // find new 'minlength2' and 'index' values for 'tosink'
      minlength2 = large;
      level2 = 0;
      for (i=1; i<=n; i++) {
        // finds minlength2 and stores in stack2[1:level2] all values
        // of k such that tosink[k] = minlength2 and tosink[i] = key2
        temp = tosink[i];
        if (temp > key2) {
          if (temp < minlength2) {
            level2 = 1;
            minlength2 = temp;
            stack2[level2] = i;
          }
          else {
            if (temp == minlength2) {
              level2++;
              stack2[level2] = i;
            }
          }
        }
      }
    }
    // compute convergence criterion
    minlength3 = large;
    for (i=1; i<=n; i++) {
      lensum = fromsource[i] + tosink[i];
      if (lensum < minlength3) {
        minlength3 = lensum;
        mnode = i;
      }
    }
  } while (minlength3 > minlength1 + minlength2);
```

```
  // two ends of a shortest path meet in 'mnode' unravel the path
  num1 = mnode;
  path[n] = mnode;
  if (mnode != source) {
    numnodes = n - 1;
    while (true) {
      num2 = parent[num1];
      if (num2 == source) break;
      num1 = num2;
      path[numnodes] = num2;
      numnodes--;
    }
  }
  else
    numnodes = n;
  path[1] = source;
  num1 = numnodes + 1;
  numnodes = 2;
  while (num1 <= n) {
    path[numnodes] = path[num1];
    numnodes++;
    num1++;
  }
  if (mnode != sink) {
    num1 = mnode;
    while (true) {
      num2 = child[num1];
      if (num2 == sink) break;
        num1 = num2;
        path[numnodes] = num2;
        numnodes++;
    }
    path[numnodes] = sink;
  }
  pathlength = fromsource[mnode] + tosink[mnode];
  path[0] = numnodes;
  return pathlength;
}
```

Example:

Find the shortest path from node 5 to node 4 in the following digraph.

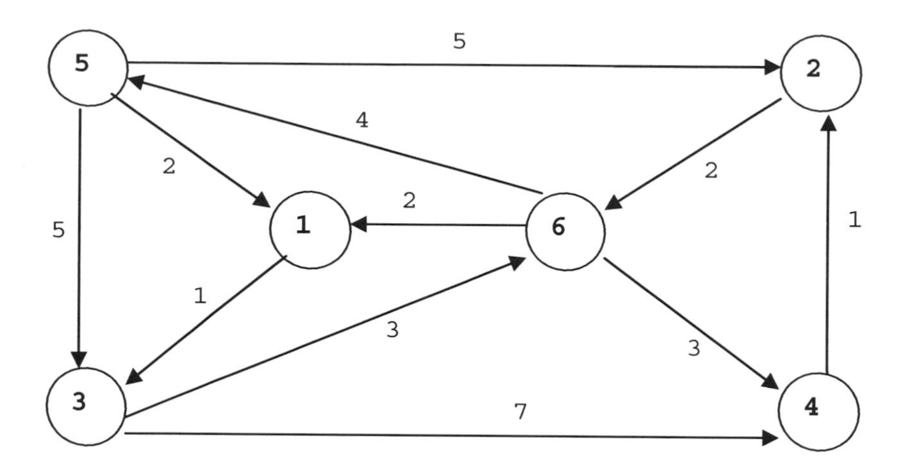


```
package GraphAlgorithms;

public class Test_oneShortestPath extends Object {

  public static void main(String args[]) {

    int k;
    int n=6, m=11;
    int source = 5, sink = 4;
    int path[] = new int[n+1];
    int nodei[] =  {0,4,6,5,3,5,6,3,6,1,2,5};
    int nodej[] =  {0,2,4,2,6,1,1,4,5,3,6,3};
    int weight[] = {0,1,3,5,3,2,2,7,4,1,2,5};

    k = GraphAlgo.oneShortestPath(n,m,nodei,nodej,weight,source,sink,path);
    System.out.println("The shortest path from node " + source +
                       " to node " + sink + " is: ");
    for (int i=1; i<=path[0]; i++)
      System.out.print("   " + path[i]);
    System.out.println("\nLength of the shortest path is " + k);
  }
}
```

Output:

```
The shortest path from node 5 to node 4 is:
  5  1  3  6  4
Length of the shortest path is 9
```

3.4 All Shortest Path Length

The following procedure [Y72] finds the length of all shortest paths from a specified source node to every other node in a given graph of n nodes with nonnegative edge lengths. A mix of directed and undirected edges are allowed in the given graph. The algorithm requires $n^3/2$ additions and n^3 comparisons.

Let d(u,v) be the length of edge (u,v) in a complete directed graph of n nodes. The length of the shortest path from node 1 to node i, S(i), i = 2,3,...,n, can be obtained as follows.

Step 1. Set p = 1, k = n, S(1) = 0, L(i) = i, S(i) = ∞, for i = 2,3,...,n.

Step 2. For i = 2,3,...,n, do the following:
- let j = L(i) and compute S(j) = min(S(j), S(p) + d(p,j));
- if the value of S(j) is less than the current minimum, say S(q), during this exection of Step 2 then set q = j and t = i.

Step 3. Set p = q, L(t) = L(k), k = k – 1. If k = 1 then stop, else return to Step 2.

Procedure parameters:

void allShortestPathLength (*n,m,nodei,nodej,directed,weight ,root,,mindistance*)

n: int;
entry: number of nodes of the graph,
nodes of the graph are labeled from 1 to *n*.

m: int;
entry: number of edges of the graph.

nodei, nodej: int[*m*+1];
entry: *nodei[p]* and *nodej[p]* are the end nodes of the p-th edge in the graph, p=1,2,...,*m*.

directed: boolean[*m*+1];
entry: *directed[p]* = true if the edge p is directed from node *nodei[p]* to node *nodej[p]*, otherwise *directed[p]* = false if the edge (*nodei[p]*, *nodej[p]*) is undirected, for p=1,2,...,*m*.

weight: int[*m*+1];
entry: *weight[p]* is the nonnegative length of edge p in the graph, p=1,2,...,*m*.

root: int;
entry: the source node of the graph.

mindistance: int[*n*+1];
exit: *mindistance[i]* is the length of the shortest path from *root* to node i, for i=1,2,...,*n*.

```
public static void allShortestPathLength(int n, int m, int nodei[],
                      int nodej[], boolean directed[], int weight[],
                      int root, int mindistance[])
{
  int i,j,k,n2,large,nodeu,nodev,minlen,temp,minj=0,minv=0;
  int location[] = new int[n+1];
```

```
int distance[][] = new int[n+1][n+1];

// obtain a large number greater than all edge weights
large = 1;
for (k=1; k<=m; k++)
  large += weight[k];

// set up the distance matrix
for (i=1; i<=n; i++)
  for (j=1; j<=n; j++)
    distance[i][j] = (i == j) ? 0 : large;
for (k=1; k<=m; k++) {
  i = nodei[k];
  j = nodej[k];
  if (directed[k])
    distance[i][j] = weight[k];
  else
    distance[i][j] = distance[j][i] = weight[k];
}

if (root != 1) {
  // interchange rows 1 and root
  for (i=1; i<=n; i++) {
    temp = distance[1][i];
    distance[1][i] = distance[root][i];
    distance[root][i] = temp;
  }
  // interchange columns 1 and root
  for (i=1; i<=n; i++) {
    temp = distance[i][1];
    distance[i][1] = distance[i][root];
    distance[i][root] = temp;
  }
}
nodeu = 1;
n2 = n + 2;
for (i=1; i<=n; i++) {
  location[i] = i;
  mindistance[i] = distance[nodeu][i];
}
for (i=2; i<=n; i++) {
  k = n2 - i;
  minlen = large;
  for (j=2; j<=k; j++) {
    nodev = location[j];
    temp = mindistance[nodeu] + distance[nodeu][nodev];
    if (temp < mindistance[nodev]) mindistance[nodev] = temp;
    if (minlen > mindistance[nodev]) {
      minlen = mindistance[nodev];
      minv = nodev;
```

```
        minj = j;
      }
    }
    nodeu = minv;
    location[minj] = location[k];
  }
  if (root != 1) {
    mindistance[1] = mindistance[root];
    mindistance[root] = 0;
    // interchange rows 1 and root
    for (i=1; i<=n; i++) {
      temp = distance[1][i];
      distance[1][i] = distance[root][i];
      distance[root][i] = temp;
    }
    // interchange columns 1 and root
    for (i=1; i<=n; i++) {
      temp = distance[i][1];
      distance[i][1] = distance[i][root];
      distance[i][root] = temp;
    }
  }
}
```

Example:

Find the shortest path distance from node 5 to every other node in the following mixed graph.

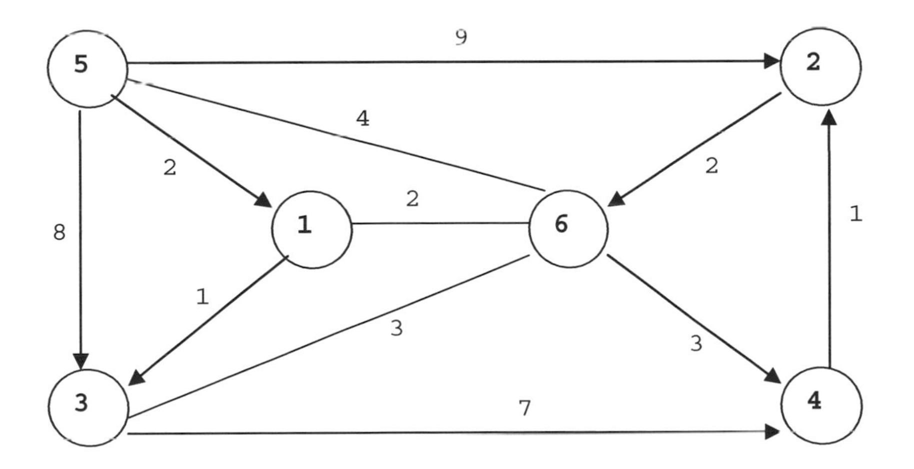

```
package GraphAlgorithms;

public class Test_allShortestPathLength extends Object {

  public static void main(String args[]) {

    int n = 6;
    int m = 11;
    int root = 5;
    int mindistance[] = new int[n+1];
    int nodei[] =  {0,4,6,5,3,5,6,3,6,1,2,5};
    int nodej[] =  {0,2,4,2,6,1,1,4,5,3,6,3};
    int weight[] = {0,1,3,9,3,2,2,7,4,1,2,8};
    boolean directed[] = {false, true, true, true, false, true,
                          false, true, false, true, true, true};

    GraphAlgo.allShortestPathLength(n,m,nodei,nodej,directed,
                                    weight,root,mindistance);
    for (int i=1; i<=n; i++)
      if (i != root)
        System.out.println("Shortest path distance from node " +
             root + " to node " + i + " is  " + mindistance[i]);
  }
}
```

Output:

```
Shortest path distance from node 5 to node 1 is  2
Shortest path distance from node 5 to node 2 is  8
Shortest path distance from node 5 to node 3 is  3
Shortest path distance from node 5 to node 4 is  7
Shortest path distance from node 5 to node 6 is  4
```

3.5 Shortest Path Tree

The following label-correcting method [F56, M57] finds the shortest paths from a specified source node to every other node in a directed graph of n nodes with lengths associated with edges. The edge lengths may be nonpositive, but no cycle of negative length is present in the graph. The running time of the algorithm is $O(n^3)$.

Let d(u,v) be the length of the edge directed from node u to node v in the given graph. The shortest paths from a specified node s to every other node can be found by the following label-correcting method.

Step 1. Let T be a tree consisting of the root s alone. Set L(s) = 0, and L(i) = ∞ for every other node i.

Step 2. Find an edge (u,v) such that L(u) + d(u,v) < L(v). Set L(v) = L(u) + d(u,v). Include the edge (u,v) in T and remove any edge in T which ends at v.

Step 3. Repeat Step 2 until L(u) + d(u,v) ≥ L(v) for all edges (u,v). The tree T will contain a shortest path from s to every other node.

Procedure parameters:

void shortestPathTree (*n, m ,nodei, nodej, weight, root, minidistance, treearc*)

n: int;
entry: number of nodes of the directed graph, nodes of the graph are labeled from 1 to *n*.

m: int;
entry: number of edges of the directed graph.

nodei, nodej: int[*m*+1];
entry: *nodei[p]* and *nodej[p]* are the end nodes of the p-th edge in the graph, p=1,2,...,*m*.

weight: int[*m*+1];
entry: *weight[p]* is the length of edge p in the graph, p=1,2,...,*m*. The length of an edge can be nonpositve, but no cycle of negative length is present in the graph.

root: int;
entry: the source node of the graph.

mindistance: int[*n*+1];
exit: *mindistance[i]* is the length of the shortest path from *root* to node i, for i=1,2,...,*n*. *mindistance[root]* is equal to zero.

treearc: int[*n*+1];
exit: The directed edge from node *treearc[i]* to node i is on the shortest path tree. *treearc[root]* is equal to zero.

```
public static void shortestPathTree(int n, int m, int nodei[], int nodej[],
                    int weight[], int root, int mindistance[], int treearc[])
{
  int i,j,k,large,nodeu,nodev,nodey,start,index,last,p,lensum,lenu;
  int queue[] = new int[n+1];
  int firstedges[] = new int[n+2];
  int endnode[] = new int[m+1];
  int origin[] = new int[m+1];
  boolean mark[] = new boolean[n+1];

  // obtain a large number greater than all edge weights
  large = 1;
  for (i=1; i<=m; i++)
    large += (weight[i] > 0) ? weight[i] : 0;

  // set up the forward star representation of the graph
  k = 0;
  for (i=1; i<=n; i++) {
    firstedges[i] = k + 1;
    for (j=1; j<=m; j++) {
```

```
      if (nodei[j] == i) {
        k++;
        origin[k] = j;
        endnode[k] = nodej[j];
      }
    }
  }
  firstedges[n+1] = m + 1;
  for (i=1; i<=n; i++) {
    treearc[i] = 0;
    mark[i] = true;
    mindistance[i] = large;
  }
  mindistance[root] = 0;
  nodev = 1;
  nodey = nodev;
  nodeu = root;
  while (true) {
    lenu = mindistance[nodeu];
    start = firstedges[nodeu];
    if (start != 0) {
      index = nodeu + 1;
      while (true) {
        last = firstedges[index] - 1;
        if (last > -1) break;
        index++;
      }
      for (i=start; i<=last; i++) {
        p = endnode[i];
        lensum = weight[origin[i]] + lenu;
        if (mindistance[p] > lensum) {
          mindistance[p] = lensum;
          treearc[p] = nodeu;
          if (mark[p]) {
            mark[p] = false;
            queue[nodey] = p;
            nodey++;
            if (nodey > n) nodey = 1;
          }
        }
      }
    }
    if (nodev == nodey) break;
    nodeu = queue[nodev];
    mark[nodeu] = true;
    nodev++;
    if (nodev > n) nodev = 1;
  }
}
```

Example:

Find the shortest path lengths from node 5 to every other node in the following digraph.

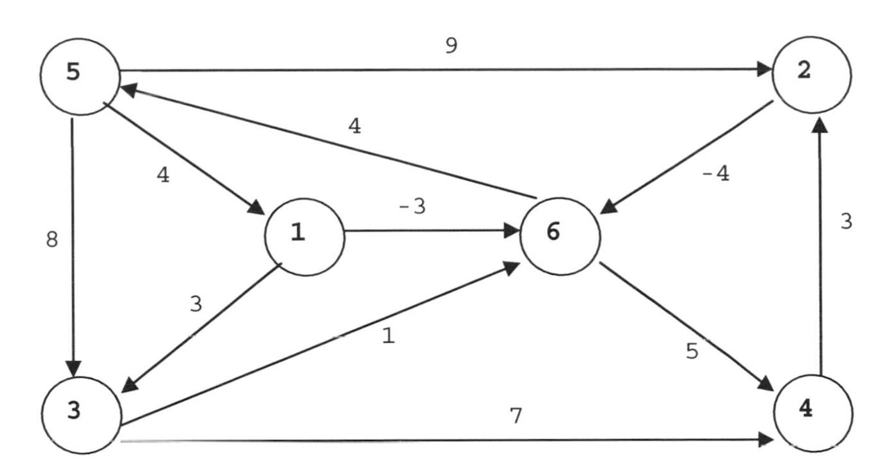


```
package GraphAlgorithms;

public class Test_shortestPathTree extends Object {

  public static void main(String args[]) {

    int n = 6;
    int m = 11;
    int root = 5;
    int treearc[] = new int[n+1];
    int mindistance[] = new int[n+1];
    int nodei[] =  {0,4,6,5,3,5, 1,3,6,1, 2,5};
    int nodej[] =  {0,2,4,2,6,1, 6,4,5,3, 6,3};
    int weight[] = {0,3,5,9,1,4,-3,7,4,3,-4,8};

    GraphAlgo.shortestPathTree(n,m,nodei,nodej,weight,root,mindistance,treearc);
    for (int i=1; i<=n; i++)
      if (i != root)
        System.out.println("Shortest path distance from node " +
             root + " to node " + i + " is  " + mindistance[i]);
    System.out.println("\nEdges in the shortest path tree are:");
    for (int i=1; i<=n; i++)
      if (i != root)
        System.out.println("  (" + treearc[i] + ", " + i + ")");
  }
}
```

Output:

```
Shortest path distance from node 5 to node 1 is  4
Shortest path distance from node 5 to node 2 is  9
Shortest path distance from node 5 to node 3 is  7
Shortest path distance from node 5 to node 4 is  6
Shortest path distance from node 5 to node 6 is  1

Edges in the shortest path tree are:
   (5, 1)
   (5, 2)
   (1, 3)
   (6, 4)
   (1, 6)
```

3.6 All Pairs Shortest Paths

The following procedure [F62] finds the shortest paths between all pairs of nodes in a mixed graph of n nodes with given edge lengths. Negative edge lengths are allowed but there should not be any negative cycle in the graph. The running time of the algorithm is $O(n^3)$.

Let d(u,v) be the length of edge from node u to node v. Let L be a sufficiently large number. It is assumed that d(u,u) = 0 for u = 1, 2, ..., n, and d(u,v) = L if there is no edge from node u to node v in the graph.

Step 1. Set k = 0.

Step 2. Set k = k + 1.

Step 3. For all i ≠ k, such that d(i,k) ≠ L, and all j ≠ k, such that d(k,j) ≠ L, compute

$$d(i,j) = \min(d(i,j),\ d(i,k) + d(k,j)).$$

Step 4. If k = n then stop; d(i,j) are the lengths of all the shortest paths for all i and j; otherwise return to Step 2.

Procedure parameters:

void allPairsShortestPaths (*n, dist, big, startnode, endnode, path*)

n: int;
entry: number of nodes of the mixed graph,
nodes of the graph are labeled from 1 to *n*.

dist: int[*n*+1][*n*+1];
entry: *dist[i][j]* is the distance of the edge between node i and node j, and *dist[i][i]* = 0, for i=1,2,...,*n* and j=1,2,...,*n*.
dist[i][j] = *big* if there is no edge from node i to node j.
The edges can be directed or undirected. The edge distance can be negative, but there should not be any negative length cycle in the graph.

exit: *dist[i][j]* is the length of the shortest path from node i to node j, for i=1,2,...,*n* and j=1,2,...,*n*.

big: int;
entry: A sufficiently large number greater than the sum of all edge lengths of the graph.

startnode, endnode: int;
entry: the shortest path from *startnode* to *endnode* is requested.
If *startnode* = 0 then no specific shortest path is requested; the output array *path* will be ignored.

path: int[*n*+1];
exit: *path[0]* is the total number of nodes of the shortest path from *startnode* to *endnode*, and the shortest path is given by
path[1], path[2], ..., path[path[0]].

```
public static void allPairsShortestPaths(int n, int dist[][], int big,
                                int startnode, int endnode, int path[])
{
  int i,j,k,d,num,node;
  int next[][] = new int[n+1][n+1];
  int order[] = new int[n+1];

  // compute the shortest path distance matrix
  for (i=1; i<=n; i++)
    for (j=1; j<=n; j++)
      next[i][j] = i;
  for (i=1; i<=n; i++)
    for (j=1; j<=n; j++)
      if (dist[j][i] < big)
        for (k=1; k<=n; k++)
          if (dist[i][k] < big) {
            d = dist[j][i] + dist[i][k];
            if (d < dist[j][k]) {
               dist[j][k] = d;
               next[j][k] = next[i][k];
            }
          }
  // find the shortest path from startnode to endnode
  if (startnode == 0) return;
  j = endnode;
  num = 1;
  order[num] = endnode;
  while (true) {
    node = next[startnode][j];
    num++;
    order[num] = node;
    if (node == startnode) break;
    j = node;
  }
  for (i=1; i<=num; i++)
    path[i] = order[num-i+1];
```

```
  path[0] = num;
}
```

Example:

Find the shortest path distance matrix and the shortest path from node 5 to node 2 in the following mixed graph.

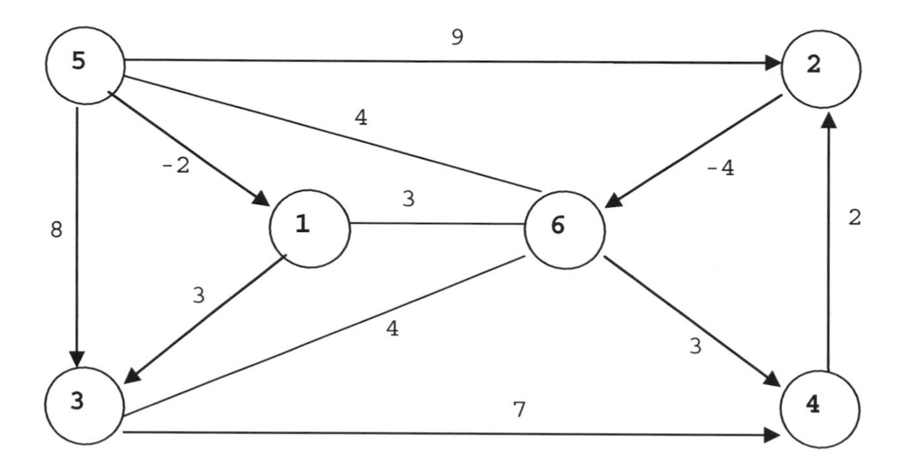


```
package GraphAlgorithms;

public class Test_allPairsShortestPaths extends Object {

  public static void main(String args[]) {

    int startnode,endnode;
    int n=6;
    int big = 99;
    int path[] = new int[n+1];
    int dist[][] = {{0,  0,  0,  0,  0,  0,  0},
                    {0,  0, 99,  3, 99, 99,  3},
                    {0, 99,  0, 99, 99, 99, -4},
                    {0, 99, 99,  0,  7, 99,  4},
                    {0, 99,  2, 99,  0, 99, 99},
                    {0, -2,  9,  8, 99,  0,  4},
                    {0,  3, 99,  4,  3,  4,  0}};

    startnode = 5;
    endnode = 2;
    GraphAlgo.allPairsShortestPaths(n,dist,big,startnode,endnode,path);
    System.out.println("The shortest distance matrix between all nodes:");
    for (int i=1; i<=n; i++) {
      for (int j=1; j<=n; j++)
        System.out.printf("%4d",dist[i][j]);
      System.out.println();
```

```
    }
    System.out.print("\nShortest path from node " + startnode +
                     " to node " + endnode + " is: ");
    for (int i=1; i<=path[0]; i++)
      System.out.print("   " + path[i]);
    System.out.println();
  }
}
```

Output:

```
The shortest distance matrix between all nodes:
   0   8   3   6   7   3
  -2   0   0  -1   0  -4
   6   9   0   7   8   4
   0   2   2   0   2  -2
  -2   6   1   4   0   1
   2   5   4   3   4   0

Shortest path from node 5 to node 2 is:   5  1  6  4  2
```

3.7 k Shortest Paths

Given an integer $k > 0$ and a specified source node, the problem is to find k shortest paths from a specified source node to every other node in a directed graph of n nodes with lengths associated with edges. The edge lengths may be nonpositive, but no cycle of negative length is present in the graph. Repeated nodes are allowed in each of the shortest paths. The running time of the algorithm [S76] is $O(n^2k^2\log n)$.

Let d(u,v) be the length of the edge directed from node u to node v in the given graph. The k shortest paths problem is solved by a label-correcting method in which an initial guess is given to the k shortest path lengths. Then the tentative k shortest path lengths will be improved successively. A sequence of forward and backward iterations will be employed. During the forward iteration, the nodes are examined in the order 1,2,...,n, and only edges (i,j) with i<j are processed. During the backward iteration, the nodes are examined in the order n,n−1, ...,1, and only edges (i,j) with i>j are processed. The alternating forward and backward iterations are continued until the node labels at two consecutive iterations coincide, in which case no improvement is possible. The algorithm can be outlined as follows.

Step 1. Initialize the required k shortest path lengths, a k-vector

$S(i) = (s(i,1), s(i,2), \ldots, s(i,k))$,

from node 1 to node i by setting

$S(1) = (0, \infty, \ldots, \infty)$, and

$S(i) = (\infty,\infty, \ldots, \infty)$ for $i = 2,3,\ldots,n$.

Step 2. For j = 2 to n, do the following:
for every node i adjacent to node j, where $i < j$, if
$\{s(i,p) + d(i,j): p = 1,2,\ldots,k\}$
gives a smaller path length than any one of the tentative k shortest path lengths in S(j) then the current k-vector S(j) is updated by including this smaller path length.

Step 3. If none of the node labels S(j) has changed in Step 2 then go to Step 6.

Step 4. For j = n−1 to 1, do the following:
for every node i adjacent to node j, where $i > j$, process the edge (i,j) in the same way as in Step 2.

Step 5. If none of the node labels S(j) has changed in Step 4 then go to Step 6, otherwise return to Step 2.

Step 6. The k shortest path lengths have been found. The paths can be reconstructed from the path length information. If a j-th shortest path P of length t from node u to node v passes through w, then the subpath of P extending from node u to node w is an i-th shortest path, for some i, $1 \le i \le j$. This can be used to determine the penultimate node w on a j-th shortest path of known length t from node u to node v. This backtracking method can be applied repeatedly to produce all the k shortest paths from node u to node v.

Procedure parameters:

void kShortestPaths(*n,m ,nodei,nodej,weight,k,source,sink,pathnodes,dist,path*)

n: int;
entry: number of nodes of the directed graph,
nodes of the graph are labeled from 1 to *n*.

m: int;
entry: number of edges of the directed graph.

nodei, nodej: int[*m*+1];
entry: *nodei[p]* and *nodej[p]* are the end nodes of the p-th edge in the graph, p=1,2,...,*m*.
exit: the list of edges (*nodei[p], nodej[p]*), p=1,2,...,*m*, will be reordered such that the array *nodej* is sorted in nondecreasing order.

weight: int[*m*+1];
entry: *weight[p]* is the length of edge p in the graph, p=1,2,...,*m*. The length of an edge can be nonpositve, but no cycle of negative length is present in the graph.
exit: the array *weight* will be reordered such that the array *nodej* is sorted in nondecreasing order.

k, source, sink: int;
entry: the *k* shortest paths from the *source* to the *sink* are required.

pathnodes: int[2];
entry: *pathnodes[0]* is the maximum number of nodes in a possible shortest path which allows repeated nodes.
pathnodes[1] is the maximum number of shortest paths to be generated from the *source* to the *sink*.
exit: If there is no path from the *source* to the *sink* then *pathnodes[1]* will be set to zero.

If the number of nodes in a shortest path exceeds the input value of *pathnodes[0]* then *pathnodes[0]* will be set equal to zero. In this case, run the procedure again by using a larger input value of *pathnodes[0]*.

dist: int[*n*+1][*k*+1];

exit: *dist[i][j]* is the length of the j-th shortest path from the *source* to node i, for i=1,2,...,*n*, and j=1,2,...,*k*. If node i is not reachable from source in the j-th shortest path then *dist[i][j]* will be set equal to a number greater than the sum of all edge lengths in the graph.

path: int[*pathnodes[1]*+1][*pathnodes[0]*+2];

exit: *path[i][0]* is the number of nodes in the i-th shortest path from the *source* to the *sink*. The nodes of the i-th shortest path are:

path[i][1], path[i][2], ..., path[i][path[i][0]]

for i = 1, 2, ..., *pathnodes[1]*.

```
public static void kShortestPaths(int n, int m, int nodei[], int nodej[],
                                  int weight[], int k, int source, int sink,
                                  int pathnodes[], int dist[][], int path[][])
{
  int i,j,init,iter,numforward,numbackward,large,nodeu,nodev;
  int ids1,ids2,max,index,index1,index2,index3;
  int length1,length2,length3,loop1,loop2,len,sub1=0,sub2=0;
  int elm,elmnodei,elmweight,h,p,temp;
  int arcforward[] = new int[n+1];
  int arcbackward[] = new int[n+1];
  int aux[] = new int[k+1];
  boolean best,skip;

  // use heapsort to sort the nodej array in ascending order
  // initially nodes m/2 to 1 are leaves in the heap
  for (i=m/2; i>=1; i--) {
    elm = nodej[i];
    elmnodei = nodei[i];
    elmweight = weight[i];
    index = i;
    while (true) {
      p = 2 * index;
      if (m < p) break;
      if (p + 1 <= m)
        if (nodej[p] < nodej[p+1]) p++;
      if (nodej[p] <= elm) break;
      nodej[index] = nodej[p];
      nodei[index] = nodei[p];
      weight[index] = weight[p];
      index = p;
    }
    nodej[index] = elm;
    nodei[index] = elmnodei;
    weight[index] = elmweight;
  }
```

```
// swap nodej[1] with nodej[m]
temp = nodej[1];
nodej[1] = nodej[m];
nodej[m] = temp;
temp = nodei[1];
nodei[1] = nodei[m];
nodei[m] = temp;
temp = weight[1];
weight[1] = weight[m];
weight[m] = temp;
// repeat delete the root from the heap
for (h=m-1; h>=2; h--) {
  // restore the heap structure of nodej[1] through nodej[h]
  for (i=h/2; i>=1; i--) {
    elm = nodej[i];
    elmnodei = nodei[i];
    elmweight = weight[i];
    index = i;
    while (true) {
      p = 2 * index;
      if (h < p) break;
      if (p + 1 <= h)
        if (nodej[p] < nodej[p+1]) p++;
      if (nodej[p] <= elm) break;
      nodej[index] = nodej[p];
      nodei[index] = nodei[p];
      weight[index] = weight[p];
      index = p;
    }
    nodej[index] = elm;
    nodei[index] = elmnodei;
    weight[index] = elmweight;
  }
  // swap nodej[1] and nodej[h]
  temp = nodej[1];
  nodej[1] = nodej[h];
  nodej[h] = temp;
  temp = nodei[1];
  nodei[1] = nodei[h];
  nodei[h] = temp;
  temp = weight[1];
  weight[1] = weight[h];
  weight[h] = temp;
}
// finish the heap sort
init = 0;
numforward = 0;
numbackward = 0;
for (i=1; i<=m; i++) {
  if (nodei[i] > nodej[i])
```

```
        numbackward++;
      else
        numforward++;
    }
    int adjbackward[] = new int[numbackward+1];
    int lenbackward[] = new int[numbackward+1];
    int adjforward[] = new int[numforward+1];
    int lenforward[] = new int[numforward+1];
    numforward = 0;
    numbackward = 0;
    index = 0;
    large = 1;
    for (i=1; i<=m; i++) {
      nodev = nodei[i];
      nodeu = nodej[i];
      len   = weight[i];
      large += (len > 0) ? len : 0;
      if (nodeu != index) {
        if (nodeu != index+1)
          for (j=index+1; j<=nodeu-1; j++) {
            arcbackward[j] = 0;
            arcforward[j] = 0;
          }
        if (init != 0) {
          arcforward[index] = sub1;
          arcbackward[index] = sub2;
        }
        sub1 = 0;
        sub2 = 0;
        index = nodeu;
      }
      init++;
      if (nodev <= nodeu) {
        numforward++;
        adjforward[numforward] = nodev;
        lenforward[numforward] = len;
        sub2++;
      }
      else {
        numbackward++;
        adjbackward[numbackward] = nodev;
        lenbackward[numbackward] = len;
        sub1++;
      }
    }
    arcbackward[index] = sub2;
    arcforward[index] = sub1;
    for (i=1; i<=n; i++)
      for (j=1; j<=k; j++)
        dist[i][j] = large;
```

```
dist[source][1] = 0;
iter = 1;
while (true) {
  ids2 = numbackward;
  best = true;
  i = n - 1;
  while (i > 0) {
    if (arcforward[i] != 0) {
      ids1 = ids2 - arcforward[i] + 1;
      // matrix multiplication with dist using the lower
      // triangular part of the edge distance matrix
      for (j=1; j<=k; j++)
        aux[j] = dist[i][j];
      max = aux[k];
      for (loop1=ids1; loop1<=ids2; loop1++) {
        index1 = adjbackward[loop1];
        length3 = lenbackward[loop1];
        for (loop2=1; loop2<=k; loop2++) {
          length1 = dist[index1][loop2];
          if (length1 >= large) break;
          length2 = length1 + length3;
          if (length2 >= max) break;
          j = k;
          skip = false;
          while (true) {
            if (j >= 2) {
              if (length2 < aux[j-1])
                j--;
              else {
                if (length2 == aux[j-1]) skip = true;
                break;
              }
            }
            else {
              j = 1;
              break;
            }
          }
          if (skip) continue;
          index2 = k;
          while (index2 > j) {
            aux[index2] = aux[index2 - 1];
            index2--;
          }
          aux[j] = length2;
          best = false;
          max = aux[k];
        }
      }
      if (!best)
```

```
        for (j=1; j<=k; j++)
          dist[i][j] = aux[j];
      ids2 = ids1 - 1;
    }
    i--;
  }
  if (iter != 1)
    if (best) break;
  iter++;
  ids1 = 1;
  best = true;
  for (i=2; i<=n; i++)
    if (arcbackward[i] != 0) {
      ids2 = ids1 + arcbackward[i] - 1;
      // matrix multiplication with dist using the upper
      // triangular part of the edge distance matrix
      for (j=1; j<=k; j++)
        aux[j] = dist[i][j];
      max = aux[k];
      for (loop1=ids1; loop1<=ids2; loop1++) {
        index1 = adjforward[loop1];
        length3 = lenforward[loop1];
        for (loop2=1; loop2<=k; loop2++) {
          length1 = dist[index1][loop2];
          if (length1 >= large) break;
          length2 = length1 + length3;
          if (length2 >= max) break;
          j = k;
          skip = false;
          while (true) {
            if (j >= 2) {
              if (length2 < aux[j-1])
                j--;
              else {
                if (length2 == aux[j-1]) skip = true;
                break;
              }
            }
            else {
              j = 1;
              break;
            }
          }
          if (skip) continue;
          index2 = k;
          while (index2 > j) {
            aux[index2] = aux[index2 - 1];
            index2--;
          }
          aux[j] = length2;
```

```
          best = false;
          max = aux[k];
        }
      }
      if (!best)
        for (j=1; j<=k; j++)
          dist[i][j] = aux[j];
      ids1 = ids2 + 1;
    }
  if (!best) iter++;
}
// store at most maxpth number of k shortest paths
// from source to sink, allowing repeated nodes
int ptag,numpth,numpath,up,npmax,nt,nd;
int maxpth,isub,jlen,lt;
npmax = pathnodes[0];
maxpth = pathnodes[1];
int trailnode[] = new int[n+2];
int neighbors[] = new int[m+1];
int edgelength[] = new int[m+1];
int currentnode[] = new int[npmax+1];
int position[] = new int[npmax+1];
int pathlength[] = new int[npmax+1];
boolean nextp;

init = 0;
index = 0;
large = 1;
for (i=1; i<=m; i++) {
  large += weight[i];
  nodev = nodei[i];
  nodeu = nodej[i];
  len = weight[i];
  if (nodeu != index) {
    if (nodeu != index+1)
      for (j=index+1; j<=nodeu-1; j++)
        trailnode[j] = 0;
    trailnode[nodeu] = init + 1;
    index = nodeu;
  }
  init++;
  neighbors[init] = nodev;
  edgelength[init] = len;
}
trailnode[index + 1] = init + 1;
for (i=1; i<=npmax; i++) {
  currentnode[i] = 0;
  position[i] = 0;
  pathlength[i] = 0;
}
```

```
numpath = 1;
if (source == sink) numpath = 2;
numpth = 0;
if (dist[sink][numpath] >= large) {
  // no path exists from source to sink
  pathnodes[1] = 0;
  return;
}
do {
  ptag = 1;
  length1 = dist[sink][numpath];
  if (length1 == large) return;
  length2 = length1;
  currentnode[1] = sink;
  do {
    nt = currentnode[ptag];
    ids1 = trailnode[nt];
    nd = nt;
    while ((trailnode[nd + 1] == 0) && (nd < n))
      nd++;
    up = trailnode[nd + 1] - 1;
    sub2 = ids1 + sub1;
    nextp = false;
    while (sub2 <= up) {
      isub = neighbors[sub2];
      jlen = edgelength[sub2];
      lt = length1 - jlen;
      j = 1;
      skip = false;
      while (true) {
        if ((dist[isub][j] > lt) || (j > k)) {
          sub2++;
          skip = true;
          break;
        }
        if (dist[isub][j] >= lt) break;
        j++;
      }
      if (skip) continue;
      ptag++;
      if (ptag > npmax) {
        // number of edges in a path exceeds allocated space
        pathnodes[0] = 0;
        return;
      }
      currentnode[ptag] = isub;
      position[ptag] = sub2 - ids1 + 1;
      pathlength[ptag] = jlen;
      length1 = lt;
      if (length1 != 0) {
```

```
          sub1 = 0;
          nextp = true;
          break;
        }
        if (isub != source) {
          sub1 = 0;
          nextp = true;
          break;
        }
        // store the current shortest path
        numpth++;
        for (j=1; j<=ptag; j++)
          path[numpth][j] = currentnode[ptag-j+1];
        pathnodes[1] = numpth;
        path[numpth][0] = ptag;
        path[numpth][npmax+1] = length2;
        if (numpth >= maxpth) return;
      }
      if (!nextp) {
        sub1 = position[ptag];
        currentnode[ptag] = 0;
        length1 += pathlength[ptag];
        ptag--;
      }
    } while (ptag > 0);
    numpath++;
  } while (numpath <= k);
}
```

Example:

In the following mixed graph, find three shortest path lengths (repeated nodes allowed) from node 5 to every other node, and find at most 6 of the shortest paths from node 5 to node 4.

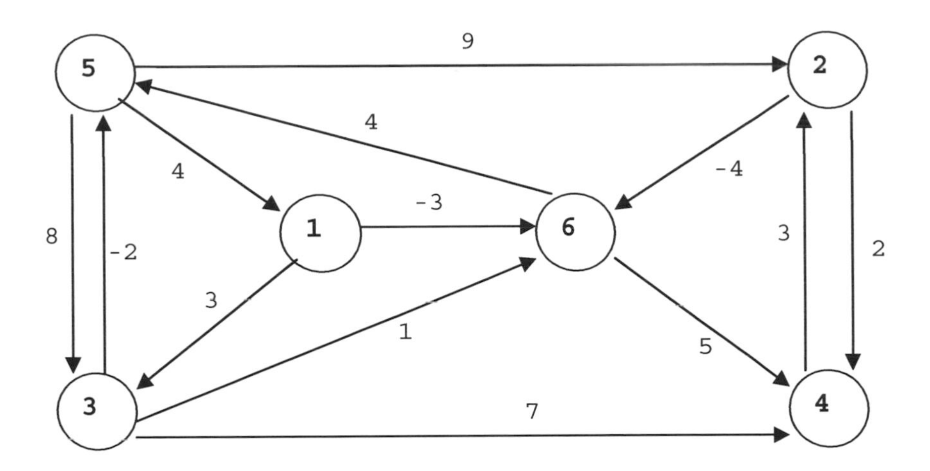


```
package GraphAlgorithms;

public class Test_kShortestPaths extends Object {

  public static void main(String args[]) {

    int pathnodes[] = {20, 6};
    int n = 6;
    int m = 13;
    int k = 3;
    int source = 5;
    int sink = 4;
    int nodei[] =  {0,2, 1,5,3,4, 2,5,6,1,6,3, 3,5};
    int nodej[] =  {0,4, 6,1,6,2, 6,2,5,3,4,4, 5,3};
    int weight[] = {0,2,-3,4,1,3,-4,9,4,3,5,7,-2,8};
    int dist[][] = new int[n+1][k+1];
    int path[][] = new int[pathnodes[1]+1][pathnodes[0]+2];

    GraphAlgo.kShortestPaths(n,m,nodei,nodej,weight,k,source,sink,
                             pathnodes,dist,path);
    System.out.println("k Shortest path lengths (k = " + k + "):");
    for (int i=1; i<=n; i++) {
      System.out.print("\nfrom node " + source + " to node " + i + ":  ");
      for (int j=1; j<=k; j++)
        System.out.printf("%4d",dist[i][j]);
    }
    if (pathnodes[0] == 0)
      System.out.println("\nNumber of arcs in a path exceeds " +
```

```
          pathnodes[0] + ". Remedy: increase size of pathnodes[0].");
    else
      if (pathnodes[1] == 0)
        System.out.println("\nThere is no path from node " + source +
                           " to node " + sink);
      else {
        System.out.print("\n\nThe following are " + pathnodes[1] +
              " shortest paths from node " + source + " to node " +
              sink + "\n       path length    Nodes in the path");
        for (int i=1; i<=pathnodes[1]; i++) {
          System.out.printf("\n %2d:      %3d          ", i,
                            path[i][pathnodes[0]+1]);
          for (int j=1; j<=path[i][0]; j++)
            System.out.printf("%3d",path[i][j]);
        }
        System.out.println();
      }
  }
}
```

Output:

```
k Shortest path lengths (k = 3):

from node 5 to node 1:     4   9  10
from node 5 to node 2:     9  13  14
from node 5 to node 3:     7   8  12
from node 5 to node 4:     6  10  11
from node 5 to node 5:     0   5   6
from node 5 to node 6:     1   5   6

The following are 6 shortest paths from node 5 to node 4
       path length    Nodes in the path
  1:        6         5  1  6  4
  2:       10         5  1  6  4  2  6  4
  3:       10         5  2  6  4
  4:       11         5  1  3  5  1  6  4
  5:       11         5  1  6  5  1  6  4
  6:       11         5  1  6  4  2  4
```

3.8 k Shortest Paths without Repeated Nodes

Given an integer $k > 0$, a specified source node and a sink node, the problem is to find k shortest paths from the source to the sink in a directed graph of n nodes with lengths associated with edges. The edge lengths may be nonpositive, but no cycle of negative length is present in the graph. Repeated nodes are not allowed in each of the shortest paths. The running time of the algorithm [Y71] is $O(kn^3)$ if all edge lengths of the input graph are nonnegative, and $O(kn^4)$ if some edge lengths are negative.

Let P(j) be the j-th shortest path from the source s to the sink t. Among the shortest paths P(1), P(2), ..., P(j-1) that have the same initial subpaths from node s to the i-th node, let Q(i,j) be the shortest of the paths that coincide with P(j−1) from node s up to the i-th node and then deviate to a node that is different from any (i+1)-th node in other paths. The algorithm can be outlined as follows.

Step 1. Find P(1), a shortest path from s to t. If there is only one shortest path then include it into a list L1. If there is more than one shortest path then include one into a list L1 and the rest into another list L2. Set j = 2.

Step 2. For each i = 1, 2, ..., j−1, find all Q(i,j) and place them into the list L2.

Step 3. Take a path from the end of the list L2. Denote this path by P(j) and move it from L2 to L1. If j = k then stop (the required list of k shortest paths is in L1), otherwise set j = j + 1 and return to Step 2.

Procedure parameters:

void kShortestPathsNoRepeatedNodes(*n,m,nodei,nodej,weight,k,source,sink,path*)

n: int;
entry: number of nodes of the directed graph, nodes of the graph are labeled from 1 to *n*.

m: int;
entry: number of edges of the directed graph.

nodei, nodej: int[*m*+1];
entry: *nodei[p]* and *nodej[p]* are the end nodes of the p-th edge in the graph, p=1,2,...,*m*.

weight: int[*m*+1];
entry: *weight[p]* is the length of edge p in the graph, p=1,2,...,*m*. The length of an edge can be nonpositve, but no cycle of negative length is present in the graph.

k,source, sink: int;
entry: the *k* shortest paths from the *source* to the *sink* are required with no repeated nodes in each path.

path: int[*k*+1][*n*+2];
exit: *path[0][0]* is the actual number of shortest paths found, where $0 \leq path[0][0] \leq k$.
path[i][n+1] is the path length of the i-th shortest path.
path[i][0] is the total number of nodes of the i-th shortest path, and the nodes of the i-th shortest path are:
path[i][1], path[i][2], ..., path[i][path[i][0]]
for i = 1, 2, ..., *path[0][0]*.

```
public static void kShortestPathsNoRepeatedNodes(int n, int m,
                int nodei[], int nodej[], int weight[], int k,
                int source, int sink, int path[][])
{
  int i,j,jj,kk,large,tail,head,node1,node2,kp3,length;
  int incrs3,index1,index2,index3,index4,ncrs4,poolc;
```

```
int j1,j2,j3,quefirstnode,linkst,treenode,examine,edge1,edge2;
int bedge,cedge,dedge,jem1,jem2,njem1,njem2,njem3,numpaths;
int upbound,lenrst,detb,deta,nrda,nrdb,quelast;
int order,nrdsize,hda,hdb,count,queorder,quefirst,quenext;
int pos1,pos2,pos3,auxda,auxdb,auxdc,low,high;
int number,nump,sub1,sub2,sub3,jsub,nextnode;
int nodep1=0,nodep2=0,nodep3=0,ncrs1=0,ncrs2=0,ncrs3=0;
int edge3=0,lenga=0,lengb=0,mark1=0,mark2=0,mshade=0;
int jump=0,incrs2=0,incrs4=0,jnd1=0,jnd2=0,jterm=0,jedge=0;
int lentab=0,nqop1=0,nqop2=0,nqsize=0,nqin=0,parm=0;
int nqout1=0,nqout2=0,nqp1=0,nqp2=0,poola=0,poolb=0;
int incrs1=0,nqfirst=0,nqlast=0;
int shortpathtree[] = new int[n+1];
int treedist[] = new int[n+1];
int arcnode[] = new int[n+1];
int arcforward[] = new int[n+1];
int arcbackward[] = new int[n+1];
int auxstore[] = new int[n+1];
int auxdist[] = new int[n+1];
int auxtree[] = new int[n+1];
int auxlink[] = new int[n+1];
int nextforward[] = new int[m+1];
int nextbackward[] = new int[m+1];
int queuefirstpath[] = new int[k+4];
int queuenextpath[] = new int[k+4];
int queuesearch[] = new int[k+4];
int kpathlength[] = new int[k+4];
int maxqueuelength = 10 * k;
int crossarc[] = new int[maxqueuelength+1];
int nextpoolentry[] = new int[maxqueuelength+1];
boolean forwrd,lastno,noroom,goon,resultfound,getpaths;
boolean loopon,lasta,lastb,rdfull,skip,force;
boolean finem1=false,finem2=false,initsp=false,nostg=false;
boolean invoke=false;

kp3 = k + 3;
// set up the network representation
for (i=1; i<=n; i++) {
  arcforward[i] = 0;
  arcbackward[i] = 0;
}
large = 1;
for (i=1; i<=m; i++) {
  large += weight[i];
  tail = nodei[i];
  head = nodej[i];
  nextforward[i] = arcforward[tail];
  arcforward[tail] = i;
  nextbackward[i] = arcbackward[head];
  arcbackward[head] = i;
```

```
}
// initialization
for (i=1; i<=n; i++)
  auxdist[i] = large;
for (i=1; i<=kp3; i++)
  queuefirstpath[i] = 0;
for (i=1; i<=maxqueuelength; i++)
  nextpoolentry[i] = i;
nextpoolentry[maxqueuelength] = 0;
// build the shortest distance tree
// treedist[i] is used to store the shortest distance of node i from
// source; shortpathtree[i] will contain the tree arc coming to node i,
// it is negative when the direction of the arc is towards source,
// it is zero if it is not reachable.
for (i=1; i<=n; i++) {
  treedist[i] = large;
  shortpathtree[i] = 0;
  arcnode[i] = 0;
}
treedist[source] = 0;
shortpathtree[source] = source;
arcnode[source] = source;
j = source;
node1 = source;
// examine neighbours of node j
do {
  edge1 = arcforward[j];
  forwrd = true;
  lastno = false;
  do {
    if (edge1 == 0)
       lastno = true;
    else {
      length = treedist[j] + weight[edge1];
      if (forwrd) {
        node2 = nodej[edge1];
        edge2 = edge1;
      }
      else {
        node2 = nodei[edge1];
        edge2 = -edge1;
      }
      if (length < treedist[node2]) {
        treedist[node2] = length;
        shortpathtree[node2] = edge2;
        if (arcnode[node2] == 0) {
          arcnode[node1] = node2;
          arcnode[node2] = node2;
          node1 = node2;
        }
```

```
        else {
          if (arcnode[node2] < 0) {
            arcnode[node2] = arcnode[j];
            arcnode[j] = node2;
            if (node1 == j) {
              node1 = node2;
              arcnode[node2] = node2;
            }
          }
        }
      }
      if (forwrd)
        edge1 = nextforward[edge1];
      else
        edge1 = nextbackward[edge1];
    }
  } while (!lastno);
  jj = j;
  j = arcnode[j];
  arcnode[jj] = -1;
} while (j != jj);
// finish building the shortest distance tree
numpaths = 0;
resultfound = false;
noroom = false;
getpaths = false;
if (shortpathtree[sink] == 0) {
  getpaths = true;
  resultfound = true;
}
if (!getpaths) {
  // initialize the storage pool
  i = 1;
  do {
    queuenextpath[i] = i;
    i++;
  } while (i <= k + 2);
  queuenextpath[k + 3] = 0;
  // initialize the priority queue
  lentab = kp3;
  low = -large;
  high = large;
  nqop1 = lentab;
  nqop2 = 0;
  nqsize = 0;
  nqin = 0;
  nqout1 = 0;
  nqout2 = 0;
  // obtain an entry from pool
  index1 = queuenextpath[1];
```

```
      queuenextpath[1] = queuenextpath[index1 + 1];
      index2 = queuenextpath[1];
      queuenextpath[1] = queuenextpath[index2 + 1];
      kpathlength[index1 + 1] = low;
      queuenextpath[index1 + 1] = index2;
      kpathlength[index2 + 1] = high;
      queuenextpath[index2 + 1] = 0;
      nqp1 = 0;
      nqp2 = 1;
      queuesearch[1] = index1;
      queuesearch[2] = index2;
      nqfirst = high;
      nqlast = low;
      // set the shortest path to the queue
      poola = queuenextpath[1];
      queuenextpath[1] = queuenextpath[poola + 1];
      poolb = poola;
      incrs1 = nextpoolentry[1];
      nextpoolentry[1] = nextpoolentry[incrs1 + 1];
      crossarc[incrs1 + 1] = shortpathtree[sink];
      nextpoolentry[incrs1 + 1] = 0;
      kpathlength[poola + 1] = treedist[sink];
      queuefirstpath[poola + 1] = incrs1;
      parm = poola;
      invoke = false;
    }
    // insert 'parm' into the priority queue
    iterate:
    while (true) {
      if (resultfound) break;
      order = kpathlength[parm + 1];
      pos1 = nqp1;
      pos2 = nqp2;
      while (pos2 - pos1 > 1) {
        pos3 = (pos1 + pos2) / 2;
        if (order > kpathlength[queuesearch[pos3 + 1] + 1])
          pos1 = pos3;
        else
          pos2 = pos3;
      }
      // linear search starting from queuesearch[pos1+1]
      index1 = queuesearch[pos1 + 1];
      do {
        index2 = index1;
        index1 = queuenextpath[index1 + 1];
      } while (kpathlength[index1 + 1] <= order);
      // insert between 'index1' and 'index2'
      queuenextpath[index2 + 1] = parm;
      queuenextpath[parm + 1] = index1;
      // update data in the queue
```

```
nqsize = nqsize + 1;
nqin = nqin + 1;
nqop1 = nqop1 - 1;
if (nqsize == 1) {
  nqfirst = order;
  nqlast = order;
}
else {
  if (order > nqlast)
    nqlast = order;
  else
    if (order < nqfirst) nqfirst = order;
}
if (nqop1 <= 0) {
  // reorganize
  index1 = queuesearch[nqp1 + 1];
  queuesearch[1] = index1;
  nqp1 = 0;
  index2 = queuesearch[nqp2 + 1];
  j3 = nqsize / lentab;
  j2 = j3 + 1;
  j1 = nqsize - ((nqsize / lentab) * lentab);
  if (j1 > 0)
    for (pos2=1; pos2<=j1; pos2++) {
      for (i=1; i<=j2; i++)
        index1 = queuenextpath[index1 + 1];
      queuesearch[pos2 + 1] = index1;
    }
  if (j3 > 0) {
    pos2 = j1 + 1;
    while (pos2 <= lentab - 1) {
      for (i=1; i<=j3; i++)
        index1 = queuenextpath[index1 + 1];
      queuesearch[pos2 + 1] = index1;
      pos2++;
    }
  }
  nqp2 = pos2;
  queuesearch[nqp2 + 1] = index2;
  nqop2 = nqop2 + 1;
  nqop1 = nqsize / 2;
  if (nqop1 < lentab) nqop1 = lentab;
}
force = false;
if (invoke) {
  if (nostg) {
    resultfound = true;
    continue iterate;
  }
  force = true;
```

```
    }
    if (!force) {
      lenga = 0;
      mark1 = 0;
      initsp = true;
      for (i=1; i<=n; i++)
        arcnode[i] = 0;
    }
    // process the next path
    while (true) {
      if (!force) {
        mark1 = mark1 + 2;
        mark2 = mark1;
        mshade = mark1 + 1;
        // obtain the first entry from the priority queue
        if (nqsize > 0) {
          index2 = queuesearch[nqp1 + 1];
          index1 = queuenextpath[index2 + 1];
          queuenextpath[index2 + 1] = queuenextpath[index1 + 1];
          nqfirst = kpathlength[queuenextpath[index1 + 1] + 1];
          if (index1 == queuesearch[nqp1 + 2]) {
            nqp1++;
            queuesearch[nqp1 + 1] = index2;
          }
          nqop1--;
          nqsize--;
          nqout1++;
          poolc = index1;
        }
        else
          poolc = 0;
        if (poolc == 0) {
          // no more paths in queue, stop
          noroom = noroom && (numpaths < k);
          resultfound = true;
          continue iterate;
        }
        queuenextpath[poolb + 1] = poolc;
        poolb = poolc;
        numpaths++;
        if (numpaths > k) {
          noroom = false;
          numpaths--;
          resultfound = true;
          continue iterate;
        }
        lengb = kpathlength[poolc + 1];
        quefirstnode = queuefirstpath[poolc + 1];
        if (lengb < lenga) {
          resultfound = true;
```

```
        continue iterate;
      }
      lenga = lengb;
      // examine the tail of the arc
      incrs2 = quefirstnode;
      ncrs2 = source;
      nodep1 = n + 1;
      // obtain data of next path
      jump = 1;
    }
    while (true) {
      if (!force) {
        // obtain data for the next path
        if (incrs2 == 0)
          linkst = 3;
        else {
          ncrs3 = ncrs2;
          incrs1 = incrs2;
          j = Math.abs(incrs1) + 1;
          incrs3 = crossarc[j];
          incrs2 = nextpoolentry[j];
          if (incrs3 > 0) {
            ncrs1 = nodei[incrs3];
            ncrs2 = nodej[incrs3];
            incrs4 = incrs3;
          }
          else {
            ncrs1 = nodej[-incrs3];
            ncrs2 = nodei[-incrs3];
            incrs4 = -incrs3;
          }
          fineml = incrs2 <= 0;
          linkst = (ncrs2 == ncrs3) ? 2 : 1;
        }
        if (jump == 1) {
          lengb -= weight[incrs4];
          nodep1--;
          auxstore[nodep1] = incrs1;
          while (ncrs1 != ncrs3) {
            j = Math.abs(shortpathtree[ncrs1]);
            lengb -= weight[j];
            ncrs1 = (shortpathtree[ncrs1] > 0) ? nodei[j] : nodej[j];
          }
          if (!fineml) {
            jump = 1;
            continue;
          }
          // store the tail of the arc
          nodep2 = nodep1;
          finem2 = fineml;
```

```
      // obtain data of next path
      jump = 2;
      continue;
    }
    if (jump == 2) {
      if (linkst == 2) {
        nodep2--;
        auxstore[nodep2] = incrs4;
        weight[incrs4] += large;
        finem2 = finem1;
        // obtain data of next path
        jump = 2;
        continue;
      }
      // close the arc on the shortest path
      finem2 = finem2 && (linkst != 3);
      if (finem2) {
        edge3 = Math.abs(shortpathtree[ncrs3]);
        nodep2--;
        auxstore[nodep2] = edge3;
        weight[edge3] += large;
      }
    }
    if (jump == 3) {
      if (linkst == 1) {
        arcnode[ncrs2] = mark2;
        while (ncrs1 != ncrs3) {
          arcnode[ncrs1] = mark2;
          if (shortpathtree[ncrs1] > 0)
            ncrs1 = nodei[shortpathtree[ncrs1]];
          else
            ncrs1 = nodej[-shortpathtree[ncrs1]];
        }
        jump = 3;
        continue;
      }
      if (linkst == 2) {
        jump = 4;
        continue;
      }
    }
    if (jump == 4) {
      if (linkst == 2) {
        jump = 4;
        continue;
      }
    }
    // mark more nodes
    if (linkst != 3) {
      arcnode[ncrs2] = mark2;
```

```
        while (ncrs1 != ncrs3) {
          arcnode[ncrs1] = mark2;
          if (shortpathtree[ncrs1] > 0)
            ncrs1 = nodei[shortpathtree[ncrs1]];
          else
            ncrs1 = nodej[-shortpathtree[ncrs1]];
        }
        jump = 3;
        continue;
      }
      // generate descendants of the tail of the arc
      nodep3 = nodep1;
      incrs1 = auxstore[nodep3];
      jnd1 = crossarc[incrs1 + 1];
      // obtain the first node of the arc traversing forward
      jnd2 = (jnd1 < 0) ? nodei[-jnd1] : nodej[jnd1];
    }
    // process a section
    do {
      if (!force) {
        nodep3++;
        jterm = jnd2;
        jedge = jnd1;
        if (nodep3 > n)
          jnd2 = source;
        else {
          incrs2 = auxstore[nodep3];
          jnd1 = crossarc[incrs2 + 1];
          jnd2 = (-jnd1 > 0) ? nodei[-jnd1] : nodej[jnd1];
        }
      }
      // process a node
      do {
        if (!force) {
          mark1 += 2;
          treenode = mark1;
          examine = mark1 + 1;
          edge3 = Math.abs(jedge);
          weight[edge3] += large;
          if (initsp) initsp = (nqin < k);
          upbound = (initsp) ? large : nqlast;
          // obtain the restricted shortest path from source to jterm
          lenrst = upbound;
          bedge = 0;
          auxdist[jterm] = 0;
          auxtree[jterm] = 0;
          auxlink[jterm] = 0;
          jem1 = jterm;
          jem2 = jem1;
          // examine next node
```

```
        do {
          njem1 = jem1;
          auxda = auxdist[njem1];
          jem1 = auxlink[njem1];
          arcnode[njem1] = treenode;
          if (auxda + treedist[njem1] + lengb >= lenrst) continue;
          goon = true;
          lasta = false;
          edge1 = arcbackward[njem1];
          // loop through arcs from njem1
          do {
            if (edge1 == 0)
              lasta = true;
            else {
              // process the arc edge1
              auxdb = auxda + weight[edge1];
              if (goon) {
                njem2 = nodei[edge1];
                edge2 = edge1;
                edge1 = nextbackward[edge1];
              }
              else {
                njem2 = nodej[edge1];
                edge2 = -edge1;
                edge1 = nextforward[edge1];
              }
              if (arcnode[njem2] != mark2) {
                auxdc = auxdb + lengb + treedist[njem2];
                if (auxdc >= lenrst) continue;
                if (arcnode[njem2] < mark2) {
                  if (shortpathtree[njem2] + edge2 == 0) {
                    arcnode[njem2] = mshade;
                  }
                  else {
                    // examine the status of the path
                    loopon = true;
                    njem3 = njem2;
                    while (loopon && (njem3 != source)) {
                      if (arcnode[njem3] < mark2) {
                        j = shortpathtree[njem3];
                        njem3 = (j > 0) ? nodei[j] : nodej[-j];
                      }
                      else
                        loopon = false;
                    }
                    if (loopon) {
                      // better path found
                      lenrst = auxdc;
                      bedge = edge2;
                      continue;
```

```
              }
              else {
                njem3 = njem2;
                lastb = false;
                do {
                  if (arcnode[njem3] < mark2) {
                    arcnode[njem3] = mshade;
                    j = shortpathtree[njem3];
                    njem3 = (j > 0) ? nodei[j] : nodej[-j];
                  }
                  else
                    lastb = true;
                } while (!lastb);
              }
            }
          }
          if ((arcnode[njem2] < treenode) ||
                        (auxdb < auxdist[njem2])) {
            // update node njem2
            auxdist[njem2] = auxdb;
            auxtree[njem2] = edge2;
            if (arcnode[njem2] != examine) {
              arcnode[njem2] = examine;
              if (jem1 == 0) {
                jem1 = njem2;
                jem2 = njem2;
                auxlink[njem2] = 0;
              }
              else {
                if (arcnode[njem2] == treenode) {
                  auxlink[njem2] = jem1;
                  jem1 = njem2;
                }
                else {
                  auxlink[njem2] = 0;
                  auxlink[jem2] = njem2;
                  jem2 = njem2;
                }
              }
            }
          }
        }
      }
    } while (!lasta);
  } while (jem1 > 0);
  arcnode[jterm] = mark2;
  // finish processing the restricted path
  if ((bedge != 0) && (lenrst < upbound)) {
    detb = 0;
    cedge = bedge;
```

```
do {
  dedge = (cedge > 0) ? nodej[cedge] : nodei[-cedge];
  if ((cedge != shortpathtree[dedge]) || (dedge == jterm)) {
    detb++;
    auxstore[detb] = cedge;
  }
  cedge = auxtree[dedge];
} while (cedge != 0);
// restore the path data
deta = detb;
nrda = nextpoolentry[1];
quelast = large;
nostg = false;
while ((deta > 0) && (nrda > 0)) {
  deta--;
  nrda = nextpoolentry[nrda + 1];
}
rdfull = (!initsp) && (numpaths + nqsize >= k);
skip = false;
while (rdfull || (deta > 0)) {
  // remove the last path from the queue
  quelast = nqlast;
  noroom = true;
  rdfull = false;
  // get the last entry from the priority queue
  if (nqsize > 0) {
    index4 = queuesearch[nqp2 + 1];
    index3 = queuesearch[nqp2];
    if (queuenextpath[index3 + 1] == index4) {
      nqp2--;
      queuesearch[nqp2 + 1] = index4;
      index3 = queuesearch[nqp2];
    }
    index2 = index3;
    while (index3 != index4) {
      index1 = index2;
      index2 = index3;
      index3 = queuenextpath[index3 + 1];
    }
    queuenextpath[index1 + 1] = index4;
    nqlast = kpathlength[index1 + 1];
    nqop1--;
    nrdsize = index2;
    nqsize--;
    nqout2++;
  }
  else
    nrdsize = 0;
  if (nrdsize == 0) {
    nostg = true;
```

```
        if (nostg) {
          resultfound = true;
          continue iterate;
        }
        skip = true;
        break;
      }
      nrda = queuefirstpath[nrdsize + 1];
      while (nrda > 0) {
        j = nrda + 1;
        deta--;
        nrdb = nrda;
        nrda = nextpoolentry[j];
        nextpoolentry[j] = nextpoolentry[1];
        nextpoolentry[1] = nrdb;
      }
      // put the entry nrdsize to pool
      queuenextpath[nrdsize + 1] = queuenextpath[1];
      queuenextpath[1] = nrdsize;
    }
    if (!skip) {
      // build the entries of crossarc and nextpoolentry
      if (lenrst >= quelast) {
        if (nostg) {
          resultfound = true;
          continue iterate;
        }
      }
      else {
        nrdb = -incrs1;
        deta = detb;
        while (deta > 0) {
          nrda = nextpoolentry[1];
          nextpoolentry[1] = nextpoolentry[nrda + 1];
          crossarc[nrda + 1] = auxstore[deta];
          nextpoolentry[nrda + 1] = nrdb;
          nrdb = nrda;
          deta--;
        }
        // obtain the entry nrdsize from pool
        nrdsize = queuenextpath[1];
        queuenextpath[1] = queuenextpath[nrdsize + 1];
        kpathlength[nrdsize + 1] = lenrst;
        queuefirstpath[nrdsize + 1] = nrdb;
        parm = nrdsize;
        invoke = true;
        continue iterate;
      }
    }
  }
```

```
          }
          force = false;
          weight[edge3] -= large;
          lengb += weight[edge3];
          if (jterm != jnd2) {
            jterm = (jedge > 0) ? nodei[jedge] : nodej[-jedge];
            jedge = shortpathtree[jterm];
          }
        } while (jterm != jnd2);
        incrs1 = incrs2;
      } while (nodep3 <= n);
      // restore the join arcs
      while (nodep2 <= nodep1 - 1) {
        j = auxstore[nodep2];
        weight[j] -= large;
        nodep2++;
      }
      // repeat with the next path
      break;
    }
  }
}
if (!getpaths) {
  // sort the paths
  hdb = poola;
  count = 0;
  do {
    hda = hdb;
    count++;
    hdb = queuenextpath[hda + 1];
    queuenextpath[hda + 1] = count;
  } while (hda != poolb);
  // release all queue entries to the pool
  j = queuesearch[nqp2 + 1];
  queuenextpath[j + 1] = queuenextpath[1];
  queuenextpath[1] = queuesearch[nqp1 + 1];
  nqp1 = 0;
  nqp2 = 0;
  hdb = 0;
  do {
    j = hdb + 1;
    hdb = queuenextpath[j];
    queuenextpath[j] = 0;
  } while (hdb != 0);
  // exchanging records
  jj = k + 2;
  for (i=1; i<=jj; i++) {
    while ((queuenextpath[i + 1] > 0) && (queuenextpath[i + 1] != i)) {
      queorder = kpathlength[i + 1];
      quefirst = queuefirstpath[i + 1];
```

```
        quenext = queuenextpath[i + 1];
        j = queuenextpath[i + 1] + 1;
        kpathlength[i + 1] = kpathlength[j];
        queuefirstpath[i + 1] = queuefirstpath[j];
        queuenextpath[i + 1] = queuenextpath[j];
        kpathlength[quenext + 1] = queorder;
        queuefirstpath[quenext + 1] = quefirst;
        queuenextpath[quenext + 1] = quenext;
      }
    }
    kpathlength[1] = source;
    queuefirstpath[1] = sink;
    queuenextpath[1] = numpaths;
  }
  // construct the edges of the k shortest paths
  for (kk=1; kk<=numpaths; kk++) {
    number = 0;
    if ((kk <= 0) || (kk > queuenextpath[1])) {
      path[kk][0] = number;
      path[0][0] = numpaths;
      return;
    }
    index2 = kpathlength[1];
    length = kpathlength[kk + 1];
    sub3 = queuefirstpath[kk + 1];
    while (sub3 != 0) {
      jsub = Math.abs(sub3) + 1;
      index3 = index2;
      if (crossarc[jsub] > 0) {
        index1 = nodei[crossarc[jsub]];
        index2 = nodej[crossarc[jsub]];
      }
      else {
        index1 = nodej[-crossarc[jsub]];
        index2 = nodei[-crossarc[jsub]];
      }
      if (index2 != index3) {
        // store the arcs
        sub2 = n;
        arcnode[sub2] = crossarc[jsub];
        while (index1 != index3) {
          sub1 = shortpathtree[index1];
          sub2--;
          if (sub2 > 0)
            arcnode[sub2] = sub1;
          else
            nump = sub1;
          index1 = (sub1 > 0) ? nodei[sub1] : nodej[-sub1];
        }
        while (sub2 <= n) {
```

```
          number++;
          arcnode[number] = arcnode[sub2];
          sub2++;
        }
      }
      sub3 = nextpoolentry[jsub];
    }
    // 'number' is the number of edges in the path
    // 'length' is the length of the path
    // 'arcnode' is the array of edge numbers of the shortest path
    nextnode = source;
    count = 0;
    for (j=1; j<=number; j++) {
      i = arcnode[j];
      count++;
      if (nodei[i] == nextnode) {
        path[kk][count] = nextnode;
        nextnode = nodej[i];
      }
      else {
        path[kk][count] = nodej[i];
        nextnode = nodei[i];
      }
    }
    count++;
    path[kk][count] = nextnode;
    path[kk][n+1] = length;
    path[kk][0] = count;
  }
  path[0][0] - numpaths;
}
```

Example:

In the following mixed graph, find four shortest node-disjoint paths from node 5 to node 4.

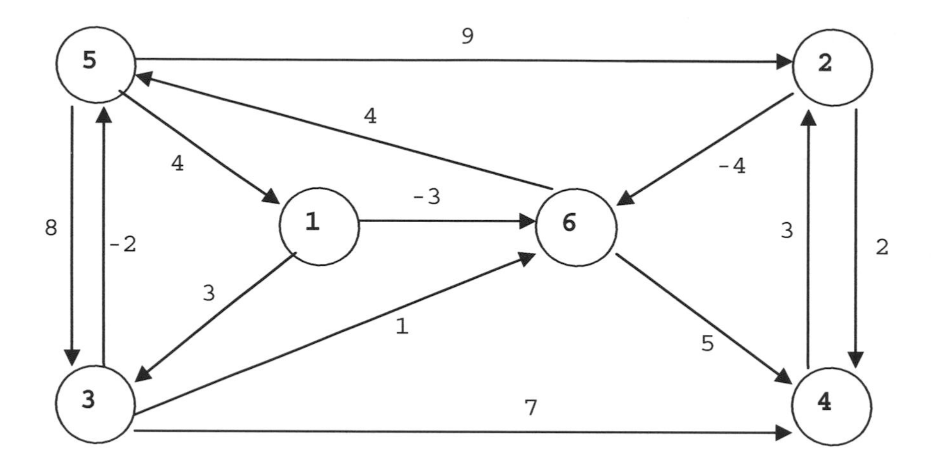


```
package GraphAlgorithms;

public class Test_kShortestPathsNoRepeatedNodes extends Object {

  public static void main(String args[]) {

    int n = 6;
    int m = 13;
    int k = 4;
    int source = 5;
    int sink = 4;
    int nodei[] =  {0,2, 1,5,3,4, 2,5,6,1,6,3, 3,5};
    int nodej[] =  {0,4, 6,1,6,2, 6,2,5,3,4,4, 5,3};
    int weight[] = {0,2,-3,4,1,3,-4,9,4,3,5,7,-2,8};

    int path[][] = new int[k+1][n+2];

    GraphAlgo.kShortestPathsNoRepeatedNodes(n,m,nodei,nodej,weight,
                                            k,source,sink,path);
    if (path[0][0] == 0)
      System.out.println("No path is found from node " +
                         source + " to node " + sink);
    else {
      System.out.print("Number of shortest paths found from node " +
                  source + " to node " + sink + " is " + path[0][0] +
                  "\nTheir shortest path lengths are ");
      for (int i=1; i<=path[0][0]; i++)
        System.out.print("  " + path[i][n+1]);
      System.out.print("\n\nThe following are " + path[0][0] +
```

```
          " shortest paths from node " + source + " to node " +
          sink + "\n        path length    Nodes in the path");
      for (int i=1; i<=path[0][0]; i++) {
        System.out.printf("\n %2d:      %3d          ",i,path[i][n+1]);
        for (int j=1; j<=path[i][0]; j++)
          System.out.printf("%3d",path[i][j]);
      }
      System.out.println();
    }
  }
}
```

Output:

```
Number of shortest paths found from node 5 to node 4 is 4
Their shortest path lengths are    6   10   11   13

The following are 4 shortest paths from node 5 to node 4
       path length     Nodes in the path
  1:        6          5  1  6  4
  2:       10          5  2  6  4
  3:       11          5  2  4
  4:       13          5  1  3  6  4
```

3.9 Euler Circuit

An *Euler circuit* of a graph is a walk which starts and ends at the same node and uses each edge exactly once. The graph can be directed or undirected. A graph is *Eulerian* if it has an Euler circuit.

A connected undirected graph is Eulerian if and only if it has no node of odd degree, and a directed graph is Eulerian if and only if the number of incoming edges is equal to the number of outgoing edges at each node. This characterization gives a straightforward procedure to decide whether a graph is Eulerian. Furthermore, an Euler circuit in an Eulerian graph G of m edges can be determined by the following $O(m)$ algorithm [L21].

Step 1. Choose any node u as the starting node. Traverse any edge (u,v) incident to node u, and then traverse any unused edge incident to node v. Repeat this process of traversing unused edges until the starting node u is reached. Let P be the resulting walk. If all edges of G are in P then stop.

Step 2. Choose any unused edge (x,y) in G such that x is in P and y is not in P. Use node x as the starting node and find another walk Q using all unused edges as in Step 1.

Step 3. Walk P and walk Q share a common node x. They can be merged to form a walk R by starting at any node s of P and then traverse P until node x is reached. Detour from P and traverse all edges of Q until node x is reached. Continue to traverse the edges of P until the starting node s is reached. Set P = R.

Step 4. Repeat Steps 2 and 3 until all edges are used.

Procedure parameters:

void EulerCircuit (*n, m, directed, nodei, nodej, trail*)

n: int;
entry: number of nodes of the graph, nodes of the graph are labeled from 1 to *n*.

m: int;
entry: number of edges of the graph.

directed: boolean;
entry: *directed* = true if the graph is directed, false otherwise.

nodei, nodej: int[*m*+1];
entry: If *directed* = false then the graph is undirected, *nodei[p]* and *nodej[p]* are the end nodes of the p-th edge in the graph.
If *directed* = true then the graph is directed, the p-th edge is directed from *nodei[p]* to *nodej[p]*, for p=1,2,...,*m*.

trail int[*m*+1];
exit: *trail[0]* = 1 if the graph is non-Eulerian.
trail[0] = 0 if the graph is Eulerian, and the edge numbers of the Euler circuit are given by:
trail[1], *trail[2]*, ..., *trail[m]*

```
public static void EulerCircuit(int n, int m, boolean directed,
                         int nodei[], int nodej[], int trail[])
{
  int i,j,k,p,index,len,traillength,stacklength;
  int endnode[] = new int[m+1];
  int stack[] = new int[m+m+1];
  boolean candidate[] = new boolean[m+1];

  // check for connectedness
  if (!connected(n,m,nodei,nodej)) {
    trail[0] = 1;
    return;
  }

  for (i=1; i<=n; i++) {
    trail[i] = 0;
    endnode[i] = 0;
  }
  if (directed) {
    // check if the directed graph is eulerian
```

```
    for (i=1; i<=m; i++) {
      j = nodei[i];
      trail[j]++;
      j = nodej[i];
      endnode[j]++;
    }
    for (i=1; i<=n; i++)
      if (trail[i] != endnode[i]) {
        trail[0] = 1;
        return;
      }
  }
  else {
    // check if the undirected graph is eulerian
    for (i=1; i<=m; i++) {
      j = nodei[i];
      endnode[j]++;
      j = nodej[i];
      endnode[j]++;
    }
    for (i=1; i<=n; i++)
      if ((endnode[i] - ((endnode[i] / 2) * 2)) != 0) {
        trail[0] = 1;
        return;
      }
  }
  // the input graph is eulerian
  trail[0] = 0;
  traillength = 1;
  stacklength = 0;
  // find the next edge
  while (true) {
    if (traillength == 1) {
      endnode[1] = nodej[1];
      stack[1] = 1;
      stack[2] = 1;
      stacklength = 2;
    }
    else {
      p = traillength - 1;
      if (traillength != 2)
        endnode[p] = nodei[trail[p]] + nodej[trail[p]] - endnode[p - 1];
      k = endnode[p];
      if (directed)
        for (i=1; i<=m; i++)
          candidate[i] = k == nodei[i];
      else
        for (i=1; i<=m; i++)
          candidate[i] = (k == nodei[i]) || (k == nodej[i]);
      for (i=1; i<=p; i++)
```

```
        candidate[trail[i]] = false;
      len = stacklength;
      for (i=1; i<=m; i++)
        if (candidate[i]) {
          len++;
          stack[len] = i;
        }
      stack[len + 1] = len - stacklength;
      stacklength = len + 1;
    }
    //  search further
    while (true) {
      index = stack[stacklength];
      stacklength--;
      if (index == 0) {
        traillength--;
        if (traillength != 0) continue;
        return;
      }
      else {
        trail[traillength] = stack[stacklength];
        stack[stacklength] = index - 1;
        if (traillength == m) return;
        traillength++;
        break;
      }
    }
  }
}
```

Example:

Find an Euler circuit in the following graph.

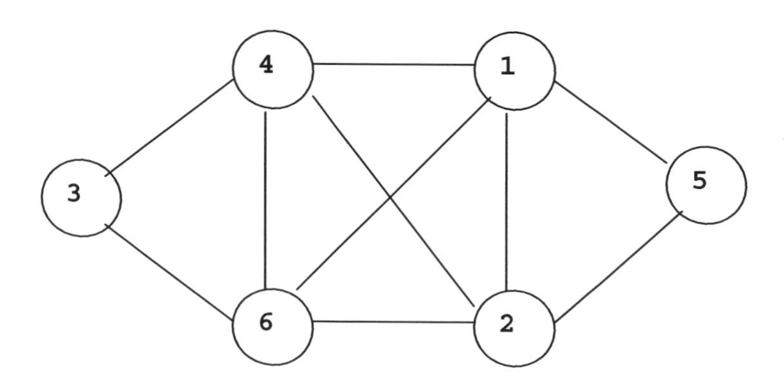

```
package GraphAlgorithms;

public class Test_EulerCircuit extends Object {

  public static void main(String args[]) {

    int n=6, m=10;
    boolean directed;
    int nodei[] = {0, 4, 1, 6, 1, 4, 1, 2, 2, 1, 3};
    int nodej[] = {0, 2, 5, 4, 2, 3, 6, 5, 6, 4, 6};
    int trail[] = new int[m+1];

    directed = false;
    GraphAlgo.EulerCircuit(n, m, directed, nodei, nodej, trail);
    if (trail[0] !- 0)
      System.out.println("Input graph is non-Eulerian");
    else {
      System.out.println("Edge numbers of the Euler circuit:");
      for (int i=1; i<=m; i++)
        System.out.print("  " + trail[i]);
      System.out.println();
    }
  }
}
```

Output:

```
Edge numbers of the Euler circuit:
  1  8  10  5  9  4  7  2  6  3
```

3.10 Hamilton Cycle

A *Hamilton Cycle* in a graph G is a cycle containing every node of G. A graph is *Hamiltonian* if it has a Hamilton cycle. In the case of a directed graph, the cycle is directed. A Hamilton cycle of a given graph G of *n* nodes will be found by an exhaustive search [W60]. Start with a single node, say node 1, as the partially constructed cycle. The cycle is grown by a backtracking procedure until a Hamilton cycle is formed. More precisely, let h(1), h(2), ..., h(k–1) be the partially constructed cycle at the k-th stage. Then the set of candidates for the next element h(k) is the set of all nodes u in G such that

if k = 1 then u = 1;

if k > 1 then (h(k–1),u) is an edge in G, and u is distinct from h(1), h(2), ..., h(k–1); furthermore, if k = n then (u,h(1)) is an edge in G and u < h(2).

The search is complete when k = n. If the set of candidates is empty at any stage then the graph is not Hamiltonian.

Procedure parameters:

void HamiltonCycle (*n, m, directed, nodei, nodej, cycle*)

n: int;
entry: number of nodes of the graph,
nodes of the graph are labeled from 1 to *n*.

m: int;
entry: number of edges of the graph.

directed: boolean;
entry: *directed* = true if the graph is directed, false otherwise.

nodei, nodej: int[*m*+1];
entry: If *directed* = false then the graph is undirected, *nodei[p]* and *nodej[p]* are the end nodes of the p-th edge in the graph.
If *directed* = true then the graph is directed, the p-th edge is directed from *nodei[p]* to *nodej[p]*, for p=1,2,...,*m*.

cycle: int[*n*+1];
exit: *cyclel[0]* = 1 if the graph is non-Hamiltonian.
cycle[0] = 0 if the graph is Hamiltonian, and the node numbers of the Hamilton cycle are given by:
cycle[1], cycle[2], ..., cycle[n]

```
public static void HamiltonCycle(int n, int m, boolean directed,
                          int nodei[], int nodej[], int cycle[])
{
  int i,j,k,stacklen,lensol,stackindex,len,len1,len2,low,up;
  int firstedges[] = new int[n+2];
  int endnode[] = new int[m+m+1];
  int stack[] = new int[m+m+1];
  boolean connect[] = new boolean[n+1];
  boolean join,skip;

  // set up the forward star representation of the graph
  k = 0;
  for (i=1; i<=n; i++) {
    firstedges[i] = k + 1;
    for (j=1; j<=m; j++) {
      if (nodei[j] == i) {
        k++;
        endnode[k] = nodej[j];
      }
      if (!directed)
        if (nodej[j] == i) {
          k++;
          endnode[k] = nodei[j];
        }
    }
  }
  firstedges[n+1] = k + 1;
  // initialize
```

```
lensol = 1;
stacklen = 0;
// find the next node
while (true) {
  if (lensol == 1) {
    stack[1] = 1;
    stack[2] = 1;
    stacklen = 2;
  }
  else {
    len1 = lensol - 1;
    len2 = cycle[len1];
    for (i=1; i<=n; i++) {
      connect[i] = false;
      low = firstedges[len2];
      up = firstedges[len2 + 1];
      if (up > low) {
        up--;
        for (k=low; k<=up; k++)
          if (endnode[k] == i) {
            connect[i] = true;
            break;
          }
      }
    }
    for (i=1; i<=len1; i++) {
      len = cycle[i];
      connect[len] = false;
    }
    len = stacklen;
    skip = false;
    if (lensol != n) {
      for (i=1; i<=n; i++)
        if (connect[i]) {
          len++;
          stack[len] = i;
        }
      stack[len + 1] = len - stacklen;
      stacklen = len + 1;
    }
    else {
      for (i=1; i<=n; i++)
        if (connect[i]) {
          if (!directed) {
            if (i > cycle[2]) {
              stack[len + 1] = len - stacklen;
              stacklen = len + 1;
              skip = true;
              break;
            }
```

```
          }
          join = false;
          low = firstedges[i];
          up = firstedges[i + 1];
          if (up > low) {
            up--;
            for (k=low; k<=up; k++)
              if (endnode[k] == 1) {
                join = true;
                break;
              }
          }
          if (join) {
            stacklen += 2;
            stack[stacklen - 1] = i;
            stack[stacklen] = 1;
          }
          else {
            stack[len + 1] = len - stacklen;
            stacklen = len + 1;
          }
          skip = true;
          break;
        }
      if (!skip) {
        stack[len + 1] = len - stacklen;
        stacklen = len + 1;
      }
    }
  }
  // search further
  while (true) {
    stackindex = stack[stacklen];
    stacklen--;
    if (stackindex == 0) {
      lensol--;
      if (lensol == 0) {
        cycle[0] = 1;
        return;
      }
      continue;
    }
    else {
      cycle[lensol] = stack[stacklen];
      stack[stacklen] = stackindex - 1;
      if (lensol == n) {
        cycle[0] = 0;
        return;
      }
      lensol++;
```

```
          break;
        }
      }
    }
}
```

Example:

Find a Hamilton cycle in the following dodecahedron.

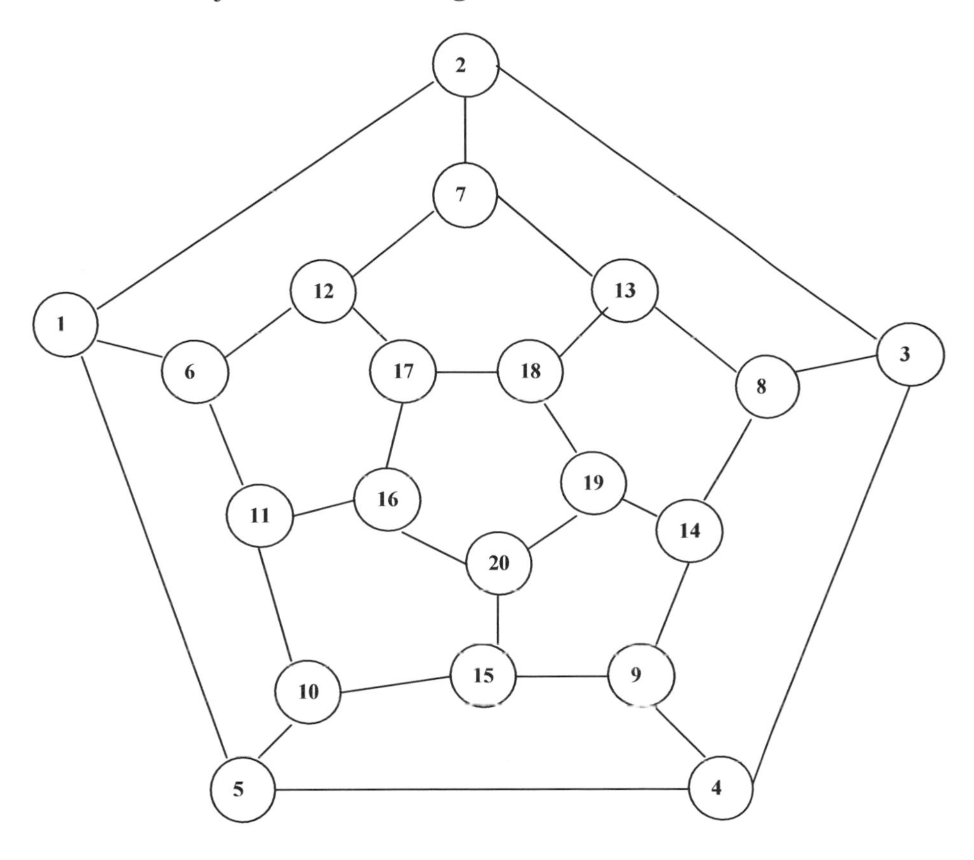


```
package GraphAlgorithms;

public class Test_HamiltonCycle extends Object {

  public static void main(String args[]) {

    int n=20, m=30;
    int cycle[] = new int[n+1];
    int nodei[] = {0,5,2,4,1,3,2,4,1,3, 5,11, 9, 7,10,13,12,
                   10,7,15,14,19,17,13,16,20,17,18,15,18,16};
    int nodej[] = {0,1,3,5,2,4,7,9,6,8,10, 6,14,12,15, 8, 6,
                   11,13, 9, 8,14,12,18,11,19,16,19,20,17,20};
    boolean directed = false;
```

```
    GraphAlgo.HamiltonCycle(n,m,directed,nodei,nodej,cycle);
    if (cycle[0] != 0)
      System.out.println("No Hamilton cycle is found.");
    else {
      System.out.println("A Hamilton cycle is found:");
      for (int i=1; i<=n; i++)
        System.out.print("  " + cycle[i]);
      System.out.println();
    }
  }
}
```

Output:

```
A Hamilton cycle is found:
  1  6  12  17  18  19  20  16  11  10  15  9  14  8  13  7  2  3  4  5
```

3.11 Chinese Postman Tour

A *postman tour* of a given connected weighted undirected graph G is a walk which contains every edge at least once. Let w(e) be the weight of edge e. The weight of the tour is the sum of the weights of all edges of the tour. The *Chinese Postman Problem* is to find the minimum weight postman tour in G. The following algorithm [EJ73, BD80] solves the problem by a labeling technique.

Note that every edge of G is visited at least once in any postman tour. Let b(e) be the number of times that edge e is in the tour. Construct a multi-graph H from G by adding b(e)−1 copies of edge e in H. After adding the copies of edges every node has an even degree. Now the postman tour in G is an Eulerian circuit in H. The Chinese Postman problem is to find the values of b(e) such that the sum of b(e)*w(e) is minimized, where $e \in G$.

Step 1. For every pair of nodes u and v of odd degree in G, find the shortest path p(u,v) connecting u and v, and let d(u,v) be the length of the shortest path.

Step 2. Consider the complete graph F whose set of nodes are all the odd degree nodes in H. Assign a weight of d(u,v) to every edge joining u and v, and solve the resulting sum matching problem (finding the largest number of edges such that no two edges are incident at the same node and the total weight is minimum).

Step 3. The edges (u,v) from the optimal matching in Step 2 represent the edges in the path p(u,v) that would have to be duplicated to obtain the Chinese postman tour.

Procedure parameters:

void ChinesePostmanTour (*n, m, startnode, nodei, nodej, cost, sol, trail*)

n: int;
entry: number of nodes of the undirected graph.
m: int;
entry: number of edges of the undirected graph.
startnode: int;
entry: starting node of the tour to be found.
nodei, nodej: int[*m*+1];
entry: *nodei[i]* and *nodej[i]* are the end nodes of the i-th edge in the graph, i=1,2,...,*m*.
cost: int[*m*+1];
entry: *cost[i]* is the cost of the i-th edge in the graph, i=1,2,...,*m*.
sol: int[*m*+1][3];
exit: the method returns an integer with the following values in *sol[0][0]*:
0: an optimal solution is found
1: the graph is connected
2: error in the input graph data
3: optimality conditions violated, need to check the input data.
If an optimal solution is found (the return value of *sol[0][0]* is zero), then *sol[1][0]* is the total cost of the optimal tour, and the total number of duplicate edges of the optimal tour is given by *sol[3][0]*; furthermore, the i-th duplicate edge is given by (*sol[i][1], sol[i][2]*), for i=1,2,..., *sol[3][0]*.
trail: int[*m*+*m*+2];
exit: If an optimal solution is found (the return value of *sol[0][0]* is zero), then the optimal tour is given by *trail[i]*, for i=1,2,..., *trail[0]*.
Note that *trail[1]* = *startnode* = *trail[trail[0]]*.

```
public static void ChinesePostmanTour(int n, int m, int startnode,
                                      int nodei[], int nodej[], int cost[],
                                      int sol[][], int trail[])
{
  int i,iplus1,j,k,idxa,idxb,idxc,idxd,idxe,wt,high,duparcs,totsolcost;
  int loch,loca,locb,locc,locd,loce,locf,locg,hub,tmpopty,tmpoptx=0;
  int nplus,p,q,cur,curnext,position=0;
  int neighbor[] = new int[m + m + 1];
  int weight[] = new int[m + m + 1];
  int degree[] = new int[n + 1];
  int next[] = new int[n + 2];
  int core[] = new int[n + 1];
  int aux1[] = new int[n + 1];
  int aux2[] = new int[n + 1];
  int aux3[] = new int[n + 1];
  int aux4[] = new int[n + 1];
  int aux5[] = new int[n + 1];
  int aux6[] = new int[n + 1];
  int tmparg[] = new int[1];
```

```
float wk1[] = new float[n + 1];
float wk2[] = new float[n + 1];
float wk3[] = new float[n + 1];
float wk4[] = new float[n + 1];
float eps,work1,work2,work3,work4;
boolean skip,complete;

eps = 0.0001f;
// check for connectedness
if (!connected(n,m,nodei,nodej)) {
  sol[0][0] = 1;
  return;
}
sol[0][0] = 0;

// store up the neighbors of each node
for (i=1; i<=n; i++)
  degree[i] = 0;
for (j=1; j<=m; j++) {
  degree[nodei[j]]++;
  degree[nodej[j]]++;
}
next[1] = 1;
for (i=1; i<=n; i++) {
  iplus1 = i + 1;
  next[iplus1] = next[i] + degree[i];
  degree[i] = 0;
}
totsolcost = 0;
high = 0;
for (j=1; j<=m; j++) {
  totsolcost += cost[j];
  k = next[nodei[j]] + degree[nodei[j]];
  neighbor[k] = nodej[j];
  weight[k] = cost[j];
  degree[nodei[j]]++;
  k = next[nodej[j]] + degree[nodej[j]];
  neighbor[k] = nodei[j];
  weight[k] = cost[j];
  degree[nodej[j]]++;
  high += cost[j];
}
nplus = n + 1;
locg = -nplus;
for (i=1; i<=n; i++)
  wk4[i] = high;
// initialization
for (p=1; p<=n; p++) {
  core[p] = p;
  aux1[p] = p;
```

```
      aux4[p] = locg;
      aux5[p] = 0;
      aux3[p] = p;
      wk1[p] = 0f;
      wk2[p] = 0f;
      i = next[p];
      loch = next[p+1];
      loca = loch - i;
      locd = loca / 2;
      locd *= 2;
      if (loca != locd) {
        loch--;
        aux4[p] = 0;
        wk3[p] = 0f;
        for (q=i; q<=loch; q++) {
          idxc = neighbor[q];
          work2 = (float) (weight[q]);
          if (wk4[idxc] > work2) {
            aux2[idxc] = p;
            wk4[idxc] = work2;
          }
        }
      }
    }
    // examine the labeling
    iterate:
    while (true) {
      work1 = high;
      for (locd=1; locd<=n; locd++)
        if (core[locd] == locd) {
          work2 = wk4[locd];
          if (aux4[locd] >= 0) {
            work2 = 0.5f * (work2 + wk3[locd]);
            if (work1 >= work2) {
              work1 = work2;
              tmpoptx = locd;
            }
          }
          else {
            if (aux5[locd] > 0) work2 += wk1[locd];
            if (work1 > work2) {
              work1 = work2;
              tmpoptx = locd;
            }
          }
        }
      work4 = ((float)high) / 2f;
      if (work1 >= work4) {
        sol[0][0] = 2;
        return;
```

```
    }
    if (aux4[tmpoptx] >= 0) {
      idxb = aux2[tmpoptx];
      idxc = aux3[tmpoptx];
      loca = core[idxb];
      locd = tmpoptx;
      loce = loca;
      while (true) {
        aux5[locd] = loce;
        idxa = aux4[locd];
        if (idxa == 0) break;
        loce = core[idxa];
        idxa = aux5[loce];
        locd = core[idxa];
      }
      hub = locd;
      locd = loca;
      loce = tmpoptx;
      while (true) {
        if (aux5[locd] > 0) break;
        aux5[locd] = loce;
        idxa = aux4[locd];
        if (idxa == 0) {
          // augmentation
          loch = 0;
          for (locb=1; locb<=n; locb++)
            if (core[locb] == locb) {
              idxd = aux4[locb];
              if (idxd >= 0) {
                if (idxd == 0) loch++;
                work2 = work1 - wk3[locb];
                wk3[locb] = 0f;
                wk1[locb] += work2;
                aux4[locb] = -idxd;
              }
              else {
                idxd = aux5[locb];
                if (idxd > 0) {
                  work2 = wk4[locb] - work1;
                  wk1[locb] += work2;
                  aux5[locb] = -idxd;
                }
              }
            }
          while (true) {
            if (locd != loca) {
              loce = aux5[locd];
              aux5[locd] = 0;
              idxd = -aux5[loce];
              idxe = aux6[loce];
```

```
      aux4[locd] = -idxe;
      idxa = -aux4[loce];
      aux4[loce] = -idxd;
      locd = core[idxa];
    }
    else {
      if (loca == tmpoptx) break;
      aux5[loca] = 0;
      aux4[loca] = -idxc;
      aux4[tmpoptx] = -idxb;
      loca = tmpoptx;
      locd = hub;
    }
  }
  aux5[tmpoptx] = 0;
  idxa = 1;
  if (loch <= 2) {
    // generate the original graph by expanding all pseudonodes
    wt = 0;
    for (locb=1; locb<=n; locb++)
      if (core[locb] == locb) {
        idxb = -aux4[locb];
        if (idxb != nplus) {
          if (idxb >= 0) {
            loca = core[idxb];
            idxc = -aux4[loca];
            tmparg[0] = position;
            cpt_DuplicateEdges(neighbor,next,idxb,idxc,tmparg);
            position = tmparg[0];
            work1 = -(float) (weight[position]);
            work1 += wk1[locb] + wk1[loca];
            work1 += wk2[idxb] + wk2[idxc];
            if (Math.abs(work1) > eps) {
              sol[0][0] = 3;
              return;
            }
            wt += weight[position];
            aux4[loca] = idxb;
            aux4[locb] = idxc;
          }
        }
      }
    for (locb=1; locb<=n; locb++) {
      while (true) {
        if (aux1[locb] == locb) break;
        hub = core[locb];
        loca = aux1[hub];
        idxb = aux5[loca];
        if (idxb > 0) {
          idxd = aux2[loca];
```

```
locd = loca;
tmparg[0] = locd;
cpt_ExpandBlossom(core,aux1,aux3,wk1,wk2,tmparg,idxd);
locd = tmparg[0];
aux1[hub] = idxd;
work3 = wk3[loca];
wk1[hub] = work3;
while (true) {
  wk2[idxd] -= work3;
  if (idxd == hub) break;
  idxd = aux1[idxd];
}
idxb = aux4[hub];
locd = core[idxb];
if (locd != hub) {
  loca = aux5[locd];
  loca = core[loca];
  idxd = aux4[locd];
  aux4[locd] = idxb;
  do {
    loce = core[idxd];
    idxb = aux5[loce];
    idxc = aux6[loce];
    locd = core[idxb];
    tmparg[0] = position;
    cpt_DuplicateEdges(neighbor,next,idxb,idxc,tmparg);
    position = tmparg[0];
    work1 = -(float)(weight[position]);
    wt += weight[position];
    work1 += wk1[locd] + wk1[loce];
    work1 += wk2[idxb] + wk2[idxc];
    if (Math.abs(work1) > eps) {
      sol[0][0] = 3;
      return;
    }
    aux4[loce] = idxc;
    idxd = aux4[locd];
    aux4[locd] = idxb;
  } while (locd != hub);
  if (loca == hub) continue;
}
while (true) {
  idxd = aux4[loca];
  locd = core[idxd];
  idxe = aux4[locd];
  tmparg[0] = position;
  cpt_DuplicateEdges(neighbor,next,idxd,idxe,tmparg);
  position = tmparg[0];
  wt += weight[position];
  work1 = -(float)(weight[position]);
```

```
        work1 += wk1[loca] + wk1[locd];
        work1 += wk2[idxd] + wk2[idxe];
        if (Math.abs(work1) > eps) {
          sol[0][0] = 3;
          return;
        }
        aux4[loca] = idxe;
        aux4[locd] = idxd;
        idxc = aux5[locd];
        loca = core[idxc];
        if (loca == hub) break;
      }
      break;
    }
    else {
      idxc = aux4[hub];
      aux1[hub] = hub;
      work3 = wk2[hub];
      wk1[hub] = 0f;
      wk2[hub] = 0f;
      do {
        idxe = aux3[loca];
        idxd = aux1[idxe];
        tmparg[0] = loca;
        cpt_ExpandBlossom(core,aux1,aux3,wk1,wk2,tmparg,idxd);
        loca = tmparg[0];
        loce = core[idxc];
        if (loce != loca) {
          idxb = aux4[loca];
          tmparg[0] = position;
          cpt_DuplicateEdges(neighbor,next,hub,idxb,tmparg);
          position = tmparg[0];
          work1 = -(float)(weight[position]);
          wt += weight[position];
          work1 += wk2[idxb] + wk1[loca] + work3;
          if (Math.abs(work1) > eps) {
            sol[0][0] = 3;
            return;
          }
        }
        else
          aux4[loca] = idxc;
        loca = idxd;
      } while (loca != hub);
    }
  }
}
// store up the duplicate edges
duparcs = 0;
i = next[2];
```

```
          for (p=2; p<=n; p++) {
            loch = next[p+1] - 1;
            for (q=i; q<=loch; q++) {
              idxd = neighbor[q];
              if (idxd <= 0) {
                idxd = -idxd;
                if (idxd <= p) {
                  duparcs++;
                  sol[duparcs][1] = p;
                  sol[duparcs][2] = idxd;
                }
              }
            }
            i = loch + 1;
          }
          cpt_Trail(n,neighbor,weight,next,aux3,core,startnode);
          // store up the optimal trail
          trail[1] = startnode;
          cur = startnode;
          curnext = 1;
          do {
            p = next[cur];
            q = aux3[cur];
            complete = true;
            for (i=q; i>=p; i--) {
              if (weight[i] > 0) {
                curnext++;
                trail[curnext] = weight[i];
                cur = weight[i];
                weight[i] = -1;
                complete = false;
                break;
              }
            }
          } while (!complete);
          trail[0] = curnext;
          sol[3][0] = duparcs;
          sol[1][0] = totsolcost + wt;
          return;
        }
        tmparg[0] = idxa;
        cpt_SecondScan(neighbor,weight,next,high,core,aux1,aux2,
                       aux3,aux4,wk1,wk2,wk3,wk4,tmparg,n);
        idxa = tmparg[0];
        continue iterate;
      }
      loce = core[idxa];
      idxa = aux5[loce];
      locd = core[idxa];
    }
```

```
    while (true) {
      if (locd == hub) {
        // shrink a blossom
        work3 = wk1[hub] + work1 - wk3[hub];
        wk1[hub] = 0f;
        idxe = hub;
        do {
          wk2[idxe] += work3;
          idxe = aux1[idxe];
        } while (idxe != hub);
        idxd = aux1[hub];
        skip = false;
        if (hub != loca) skip = true;
        do {
          if (!skip) {
            loca = tmpoptx;
            loce = aux5[hub];
          }
          skip = false;
          while (true) {
            aux1[idxe] = loce;
            idxa = -aux4[loce];
            aux4[loce] = idxa;
            wk1[loce] += wk4[loce] - work1;
            idxe = loce;
            tmparg[0] = idxe;
            cpt_ShrinkBlossom(core,aux1,wk1,wk2,hub,tmparg);
            idxe = tmparg[0];
            aux3[loce] = idxe;
            locd = core[idxa];
            aux1[idxe] = locd;
            wk1[locd] += work1 - wk3[locd];
            idxe = locd;
            tmparg[0] = idxe;
            cpt_ShrinkBlossom(core,aux1,wk1,wk2,hub,tmparg);
            idxe = tmparg[0];
            aux3[locd] = idxe;
            if (loca == locd) break;
            loce = aux5[locd];
            aux5[locd] = aux6[loce];
            aux6[locd] = aux5[loce];
          }
          if (loca == tmpoptx) {
            aux5[tmpoptx] = idxb;
            aux6[tmpoptx] = idxc;
            break;
          }
          aux5[loca] = idxc;
          aux6[loca] = idxb;
        } while (hub != tmpoptx);
```

```
          aux1[idxe] = idxd;
          loca = aux1[hub];
          aux2[loca] = idxd;
          wk3[loca] = work3;
          aux5[hub] = 0;
          wk4[hub] = high;
          wk3[hub] = work1;
          cpt_FirstScan(neighbor,weight,next,core,aux1,aux2,
                        aux3,aux4,wk1,wk2,wk3,wk4,hub);
          continue iterate;
        }
        locf = aux5[hub];
        aux5[hub] = 0;
        idxd = -aux4[locf];
        hub = core[idxd];
      }
    }
    else {
      if (aux5[tmpoptx] > 0) {
        loca = aux1[tmpoptx];
        if (loca != tmpoptx) {
          idxa = aux5[loca];
          if (idxa > 0) {
            // expand a blossom
            idxd = aux2[loca];
            locd = loca;
            tmparg[0] = locd;
            cpt_ExpandBlossom(core,aux1,aux3,wk1,wk2,tmparg,idxd);
            locd = tmparg[0];
            work3 = wk3[loca];
            wk1[tmpoptx] = work3;
            aux1[tmpoptx] = idxd;
            while (true) {
              wk2[idxd] -= work3;
              if (idxd == tmpoptx) break;
              idxd = aux1[idxd];
            }

            idxb = -aux4[tmpoptx];
            locd = core[idxb];
            idxc = aux4[locd];
            hub = core[idxc];
            if (hub != tmpoptx) {
              loce = hub;
              while (true) {
                idxa = aux5[loce];
                locd = core[idxa];
                if (locd == tmpoptx) break;
                idxa = aux4[locd];
                loce = core[idxa];
```

```
        }
        aux5[hub] = aux5[tmpoptx];
        aux5[tmpoptx] = aux6[loce];
        aux6[hub] = aux6[tmpoptx];
        aux6[tmpoptx] = idxa;
        idxd = aux4[hub];
        loca = core[idxd];
        idxe = aux4[loca];
        aux4[hub] = -idxb;
        locd = loca;
        while (true) {
          idxb = aux5[locd];
          idxc = aux6[locd];
          aux5[locd] = idxe;
          aux6[locd] = idxd;
          aux4[locd] = idxb;
          loce = core[idxb];
          idxd = aux4[loce];
          aux4[loce] = idxc;
          if (loce == tmpoptx) break;
          locd = core[idxd];
          idxe = aux4[locd];
          aux5[loce] = idxd;
          aux6[loce] = idxc;
        }
      }
      idxc = aux6[hub];
      locd = core[idxc];
      wk4[locd] = work1;
      if (locd != hub) {
        idxb = aux5[locd];
        loca = core[idxb];
        aux5[locd] = aux5[hub];
        aux6[locd] = idxc;
        do {
          idxa = aux4[locd];
          aux4[locd] = -idxa;
          loce = core[idxa];
          idxa = aux5[loce];
          aux5[loce] = -idxa;
          wk4[loce] = high;
          wk3[loce] = work1;
          locd = core[idxa];
          wk4[locd] = work1;
          cpt_FirstScan(neighbor,weight,next,core,aux1,aux2,
                        aux3,aux4,wk1,wk2,wk3,wk4,loce);
        } while (locd != hub);
        aux5[hub] = aux6[loce];
        aux6[hub] = idxa;
        if (loca == hub) continue iterate;
```

```
      }
      loce = loca;
      do {
        idxa = aux4[loce];
        aux4[loce] = -idxa;
        locd = core[idxa];
        aux5[loce] = -locd;
        idxa = aux5[locd];
        aux4[locd] = -aux4[locd];
        loce = core[idxa];
        aux5[locd] = -loce;
      } while (loce != hub);
      do {
        locd = -aux5[loca];
        tmparg[0] = loca;
        cpt_SecondScan(neighbor,weight,next,high,core,aux1,aux2,
                       aux3,aux4,wk1,wk2,wk3,wk4,tmparg,loca);
        loca = tmparg[0];
        loca = -aux5[locd];
        tmparg[0] = locd;
        cpt_SecondScan(neighbor,weight,next,high,core,aux1,aux2,
                       aux3,aux4,wk1,wk2,wk3,wk4,tmparg,locd);
        locd = tmparg[0];
      } while (loca != hub);
      continue iterate;
    }
  }
  // modify a blossom
  wk4[tmpoptx] = high;
  wk3[tmpoptx] = work1;
  i = 1;
  wk1[tmpoptx] = 0f;
  idxa = -aux4[tmpoptx];
  loca = core[idxa];
  idxb = aux4[loca];
  if (idxb == tmpoptx) {
    i = 2;
    aux4[loca] = idxa;
    idxd = aux1[tmpoptx];
    aux1[tmpoptx] = loca;
    wk1[loca] += work1 - wk3[loca];
    idxe = loca;
    tmparg[0] = idxe;
    cpt_ShrinkBlossom(core,aux1,wk1,wk2,tmpoptx,tmparg);
    idxe = tmparg[0];
    aux3[loca] = idxe;
    aux1[idxe] = idxd;
    idxb = aux6[tmpoptx];
    if (idxb == tmpoptx) {
      idxa = aux5[tmpoptx];
```

```
      loca = core[idxa];
      aux4[tmpoptx] = aux4[loca];
      aux4[loca] = idxa;
      aux5[tmpoptx] = 0;
      idxd = aux1[tmpoptx];
      aux1[tmpoptx] = loca;
      wk1[loca] += work1 - wk3[loca];
      idxe = loca;
      tmparg[0] = idxe;
      cpt_ShrinkBlossom(core,aux1,wk1,wk2,tmpoptx,tmparg);
      idxe = tmparg[0];
      aux3[loca] = idxe;
      aux1[idxe] = idxd;
      cpt_FirstScan(neighbor,weight,next,core,aux1,aux2,
                    aux3,aux4,wk1,wk2,wk3,wk4,tmpoptx);
      continue iterate;
    }
  }
  do {
    idxc = tmpoptx;
    locd = aux1[tmpoptx];
    while (true) {
      idxd = locd;
      idxe = aux3[locd];
      skip = false;
      while (true) {
        if (idxd == idxb) {
          skip = true;
          break;
        }
        if (idxd == idxe) break;
        idxd = aux1[idxd];
      }
      if (skip) break;
      locd = aux1[idxe];
      idxc = idxe;
    }
    idxd = aux1[idxe];
    aux1[idxc] = idxd;
    tmparg[0] = locd;
    cpt_ExpandBlossom(core,aux1,aux3,wk1,wk2,tmparg,idxd);
    locd = tmparg[0];
    wk4[locd] = work1;
    if (i == 2) {
      aux5[locd] = aux5[tmpoptx];
      aux6[locd] = idxb;
      aux5[tmpoptx] = 0;
      aux4[tmpoptx] = aux4[locd];
      aux4[locd] = -tmpoptx;
      cpt_FirstScan(neighbor,weight,next,core,aux1,aux2,
```

```
                   aux3,aux4,wk1,wk2,wk3,wk4,tmpoptx);
        continue iterate;
      }
      i = 2;
      aux5[locd] = tmpoptx;
      aux6[locd] = aux4[locd];
      aux4[locd] = -idxa;
      idxb = aux6[tmpoptx];
      if (idxb == tmpoptx) {
        idxa = aux5[tmpoptx];
        loca = core[idxa];
        aux4[tmpoptx] = aux4[loca];
        aux4[loca] = idxa;
        aux5[tmpoptx] = 0;
        idxd = aux1[tmpoptx];
        aux1[tmpoptx] = loca;
        wk1[loca] += work1 - wk3[loca];
        idxe = loca;
        tmparg[0] = idxe;
        cpt_ShrinkBlossom(core,aux1,wk1,wk2,tmpoptx,tmparg);
        idxe = tmparg[0];
        aux3[loca] = idxe;
        aux1[idxe] = idxd;
        cpt_FirstScan(neighbor,weight,next,core,aux1,aux2,
                   aux3,aux4,wk1,wk2,wk3,wk4,tmpoptx);
        continue iterate;
      }
    } while (core[idxb] == tmpoptx);
    aux5[locd] = aux5[tmpoptx];
    aux6[locd] = idxb;
    aux5[tmpoptx] = 0;
    locd = aux1[tmpoptx];
    if (locd == tmpoptx) {
      aux4[tmpoptx] = locg;
      tmpopty = tmpoptx;
      tmparg[0] = tmpopty;
      cpt_SecondScan(neighbor,weight,next,high,core,aux1,aux2,
                 aux3,aux4,wk1,wk2,wk3,wk4,tmparg,tmpoptx);
      tmpopty = tmparg[0];
      continue iterate;
    }
    idxe = aux3[locd];
    idxd = aux1[idxe];
    aux1[tmpoptx] = idxd;
    tmparg[0] = locd;
    cpt_ExpandBlossom(core,aux1,aux3,wk1,wk2,tmparg,idxd);
    locd = tmparg[0];
    aux4[tmpoptx] = -aux4[locd];
    aux4[locd] = -tmpoptx;
    locc = locd;
```

```
        tmparg[0] = locc;
        cpt_SecondScan(neighbor,weight,next,high,core,aux1,aux2,
                       aux3,aux4,wk1,wk2,wk3,wk4,tmparg,locd);
        locc = tmparg[0];
        tmpopty = tmpoptx;
        tmparg[0] = tmpopty;
        cpt_SecondScan(neighbor,weight,next,high,core,aux1,aux2,
                       aux3,aux4,wk1,wk2,wk3,wk4,tmparg,tmpoptx);
        tmpopty = tmparg[0];
        continue iterate;
      }
      else {
        // grow an alternating tree
        idxa = -aux4[tmpoptx];
        if (idxa <= n) {
          aux5[tmpoptx] = aux2[tmpoptx];
          aux6[tmpoptx] = aux3[tmpoptx];
          loca = core[idxa];
          aux4[loca] = -aux4[loca];
          wk4[loca] = high;
          wk3[loca] = work1;
          cpt_FirstScan(neighbor,weight,next,core,aux1,aux2,aux3,
                        aux4,wk1,wk2,wk3,wk4,loca);
          continue iterate;
        }
        else {
          idxb = aux2[tmpoptx];
          loca = core[idxb];
          aux4[tmpoptx] = aux4[loca];
          wk4[tmpoptx] = high;
          wk3[tmpoptx] = work1;
          aux4[loca] = idxb;
          wk1[loca] += work1 - wk3[loca];
          idxe = loca;
          tmparg[0] = idxe;
          cpt_ShrinkBlossom(core,aux1,wk1,wk2,tmpoptx,tmparg);
          idxe = tmparg[0];
          aux3[loca] = idxe;
          aux1[tmpoptx] = loca;
          aux1[idxe] = tmpoptx;
          cpt_FirstScan(neighbor,weight,next,core,aux1,aux2,aux3,
                        aux4,wk1,wk2,wk3,wk4,tmpoptx);
          continue iterate;
        }
      }
    }
  }
}
```

```
static private void cpt_DuplicateEdges(int neighbor[], int next[],
                                  int idxb, int idxc, int tmparg[])
{
  /* this method is used internally by ChinesePostmanTour */

  // Duplicate matching edges

  int p,q,r;

  p = tmparg[0];
  q = idxb;
  r = idxc;
  while (true) {
    p = next[q];
    while (true) {
      if (neighbor[p] == r) break;
      p++;
    }
    neighbor[p] = -r;
    if (q == idxc) break;
    q = idxc;
    r = idxb;
  }
  tmparg[0] = p;
}


static private void cpt_ExpandBlossom(int core[], int aux1[], int aux3[],
                        float wk1[], float wk2[], int tmparg[], int idxd)
{
  /* this method is used internally by ChinesePostmanTour */

  // Expanding a blossom

  int p,q,r;
  float work;

  r = tmparg[0];
  p = r;
  do {
    r = p;
    q = aux3[r];
    work = wk1[r];
    while (true) {
      core[p] = r;
      wk2[p] -= work;
      if (p == q) break;
      p = aux1[p];
    }
    p = aux1[q];
```

```
    aux1[q] = r;
  } while (p != idxd);
  tmparg[0] = r;
}


static private void cpt_FirstScan(int neighbor[], int weight[], int next[],
              int core[], int aux1[], int aux2[], int aux3[], int aux4[],
              float wk1[], float wk2[], float wk3[], float wk4[], int locb)
{
  /* this method is used internally by ChinesePostmanTour */

  // Node scanning

  int i,p,q,r,s,t,u,v;
  float work1,work2,work3,work4,work5;

  work3 = wk3[locb] - wk1[locb];
  q = locb;
  r = aux4[locb];
  t = -1;
  if (r > 0) t = core[r];
  do {
    i = next[q];
    v = next[q+1] - 1;
    work1 = wk2[q];
    for (p=i; p<=v; p++) {
      s = neighbor[p];
      u = core[s];
      if (locb != u) {
        if (t != u) {
          work4 = wk4[u];
          work2 = wk1[u] + wk2[s];
          work5 = (float)(weight[p]);
          work5 += work3 - work1 - work2;
          if (work4 > work5) {
            wk4[u] = work5;
            aux2[u] = q;
            aux3[u] = s;
          }
        }
      }
    }
    q = aux1[q];
  } while (q != locb);
}
```

```
static private void cpt_SecondScan(int neighbor[], int weight[],
       int next[], int high, int core[], int aux1[], int aux2[],
       int aux3[], int aux4[], float wk1[], float wk2[],
       float wk3[], float wk4[], int tmparg[], int v)
{
  /* this method is used internally by ChinesePostmanTour */

  // Node scanning

  int i,p,q,r,s,t,u;
  float work1,work2,work3,work4,work5;

  u = tmparg[0];
  do {
    r = core[u];
    if (r == u) {
      work4 = high;
      work2 = wk1[u];
      do {
        i = next[r];
        s = next[r+1] - 1;
        work1 = wk2[r];
        for (p=i; p<=s; p++) {
          q = neighbor[p];
          t = core[q];
          if (t != u) {
            if (aux4[t] >= 0) {
              work3 = wk3[t] - wk1[t] - wk2[q];
              work5 = (float)(weight[p]);
              work5 += work3 - work2 - work1;
              if (work4 > work5) {
                work4 = work5;
                aux2[u] = q;
                aux3[u] = r;
              }
            }
          }
        }
        r = aux1[r];
      } while (r != u);
      wk4[u] = work4;
    }
    u++;
  } while (u <= v);
  tmparg[0] = u;
}
```

```
static private void cpt_ShrinkBlossom(int core[], int aux1[],
            float wk1[], float wk2[], int locb, int tmparg[])
{
  /* this method is used internally by ChinesePostmanTour */

  // Shrinking of a blossom

  int p,q,r;
  float work;

  p = tmparg[0];
  q = p;
  work = wk1[p];
  while (true) {
    core[p] = locb;
    wk2[p] += work;
    r = aux1[p];
    if (r == q) {
      tmparg[0] = p;
      return;
    }
    p = r;
  }
}


static private void cpt_Trail(int n, int neighbor[], int weight[],
                int next[], int aux3[], int core[], int startnode)
{
  /* this method is used internally by ChinesePostmanTour */

  // Determine an Eulerian trail

  int i,nplus,p,q,r,t,u,v;
  boolean finish;

  nplus = n + 1;
  u = next[nplus];
  if (startnode <= 0 || startnode > n) startnode = 1;
  for (p=1; p<=n; p++) {
    i = next[p] - 1;
    aux3[p] = i;
    core[p] = i;
  }
  p = startnode;
  iterate:
  while (true) {
    i = core[p];
    while (true) {
      v = next[p+1] - 1;
```

```
      while (true) {
        i++;
        if (i > v) break;
        q = neighbor[i];
        if (q > n) continue;
        if (q >= 0) {
          t = core[q];
          do {
            t++;
          } while (neighbor[t] != p);
          neighbor[t] = nplus;
          t = aux3[q] + 1;
          aux3[q] = t;
          weight[t] = p;
          core[p] = i;
          p = q;
          continue iterate;
        }
        r = -p;
        q= -q;
        t = core[q];
        do {
          t++;
        } while (neighbor[t] != r);
        neighbor[t] = nplus;
        t = aux3[q] + 1;
        aux3[q] = t;
        weight[t] = p;
        t = aux3[p] + 1;
        aux3[p] = t;
        weight[t] = q;
      }
      core[p] = u;
      finish = true;
      for (p=1; p<=n; p++) {
        i = core[p];
        t = aux3[p];
        if ((t >= next[p]) && (i < u)) {
          finish = false;
          break;
        }
      }
      if (finish) return;
    }
  }
}
```

Example:

Find the Chinese postman tour of the following graph.

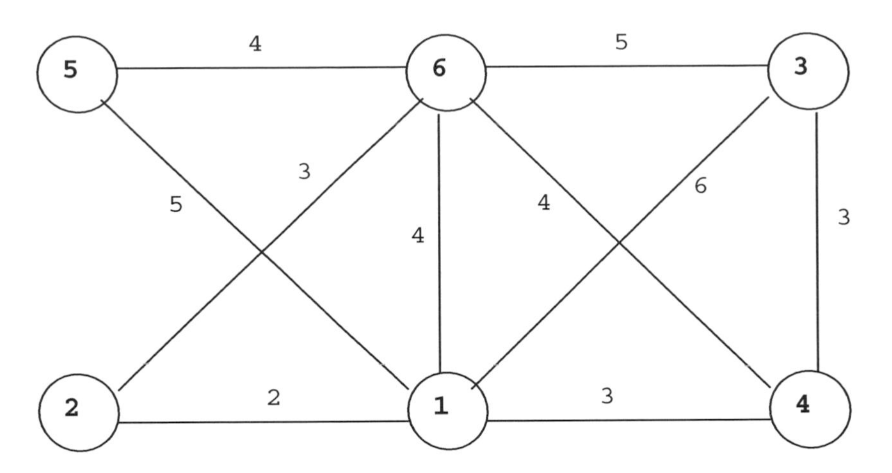


```
package GraphAlgorithms;

public class Test_ChinesePostmanTour extends Object {

  public static void main(String args[]) {

    int n=6, m=10, startnode=1;
    int sol[][] = new int[m+1][3];
    int trail[] = new int[m+m+2];
    int nodei[] = {0,3,1,1,6,1,1,1,2,4,3};
    int nodej[] = {0,6,2,3,5,5,6,4,6,6,4};
    int cost[]  = {0,5,2,6,4,5,4,3,3,4,3};

    GraphAlgo.ChinesePostmanTour(n,m,startnode,nodei,nodej,cost,sol,trail);
    if (sol[0][0] != 0)
      System.out.println("Error code returned = " + sol[0][0]);
    else {
      System.out.println("Optimal solution found.\n\nDuplicate edges:");
      for (int i=1; i<=sol[3][0]; i++)
        System.out.println(" " + sol[i][1] + " - " + sol[i][2]);
      System.out.println("\nOptimal tour:");
      for (int i=1; i<=trail[0]; i++)
        System.out.print("  " + trail[i]);
      System.out.println("\n\nOptimal tour total cost = " + sol[1][0]);
    }
  }
}
```

Output:

```
Optimal solution found.

Duplicate edges:
 4 - 3
 6 - 1

Optimal tour:
  1  4  3  4  6  1  6  5  1  3  6  2  1

Optimal tour total cost = 46
```

3.12 Traveling Salesman Problem

Given n cities and the distance between each pair of cities, the *traveling salesman problem* is to find the shortest tour for a traveling salesman to visit each city exactly once and return back to the starting city. Note that the distance matrix is not required to be symmetric. In graph theoretic terms, the problem is to find a minimum cost cycle in a given weighted complete digraph.

The problem is solved by a branch and bound recursive tree search [SDK83]. All the possible solutions are represented in a decision tree. At each step, a solution generates two branches, one contains a specific edge and the other does not contain that edge. Lower bounds are computed for each branch to help in discarding subsets of solutions. This is a well-known difficult combinatorial optimization problem and the procedure is only effective in solving small size problems.

Procedure parameters:

void travelingSalesmanProblem (*n, dist, sol*)

n: int;
entry: number of nodes in the given digraph.

dist: int[n+1][n+1];
entry: distance matrix of the given digraph. *dist[i][j]* is the distance from node i to node j, for i=1,2,...,n, j=1,2,...,n. Note that the distance matrix is not necessarily symmetric.
exit: *dist[0][0]* returns the total distance of the minimum cost cycle.

sol: int[n+1];
exit: the minimum cost cycle is given by: *sol[1], sol[2], ..., sol[n]*.

```
public static void travelingSalesmanProblem(int n, int dist[][], int sol[])
{
  int i, p;
  int row[] = new int[n + 1];
```

```
  int column[] = new int[n + 1];
  int front[] = new int[n + 1];
  int cursol[] = new int[n + 1];
  int back[] = new int[n + 1];

  for (i=1; i<=n; i++) {
    row[i] = i;
    column[i] = i;
    front[i] = 0;
    back[i] = 0;
  }
  dist[0][0] = Integer.MAX_VALUE;
  tspsearch(n, 0, 0, dist, row, column, cursol, front, back);
  p = 1;
  for (i=1; i<=n; i++) {
    sol[i] = p;
    p = cursol[p];
  }
}


static private void tspsearch(int nodes, int edges, int weight,
                              int dist[][], int row[], int column[],
                              int cursol[], int front[], int back[])
{
  /* this method is used internally by travelingSalesmanProblem*/

  int i, j, k, reduction, small, skip, stretch, candc=0, candr=0;
  int elms, head, tail, thresh, diff, miny, minx, blank;
  int cutx[] = new int[nodes + 1];
  int cuty[] = new int[nodes + 1];
  int rowvec[] = new int[nodes + 1];
  int colvec[] = new int[nodes + 1];

  elms = nodes - edges;
  reduction = 0;
  for (i=1; i<=elms ; i++) {
    small  = Integer.MAX_VALUE;
    for (j=1; j<=elms ; j++)
      small = Math.min(small, dist[row[i]][column[j]]);
    if (small > 0) {
      for (j=1; j<=elms ; j++)
        if (dist[row[i]][column[j]] < Integer.MAX_VALUE)
          dist[row[i]][column[j]] -= small;
      reduction += small;
    }
    cutx[i] = small;
  }
  for (j=1; j<=elms ; j++) {
    small = Integer.MAX_VALUE;
```

```
    for (i=1; i<=elms ; i++)
      small = Math.min(small, dist[row[i]][column[j]]);
    if (small > 0) {
      for (i=1; i<=elms ; i++)
        if (dist[row[i]][column[j]] < Integer.MAX_VALUE)
          dist[row[i]][column[j]] -= small;
      reduction += small;
    }
    cuty[j] = small;
  }
  weight += reduction;
  if (weight < dist[0][0]) {
    if (edges == (nodes - 2)) {
      for (i=1; i<=nodes; i++)
        cursol[i] = front[i];
      skip = (dist[row[1]][column[1]] == Integer.MAX_VALUE ? 1 : 2);
      cursol[row[1]] = column[3-skip];
      cursol[row[2]] = column[skip];
      dist[0][0] = weight;
    }
    else {
      diff = -Integer.MAX_VALUE;
      for (i=1; i<=elms; i++)
        for (j=1; j<=elms; j++)
          if (dist[row[i]][column[j]] == 0) {
            minx = Integer.MAX_VALUE;
            blank = 0;
            for (k=1; k<=elms; k++)
              if (dist[row[i]][column[k]] == 0)
                blank++;
              else
                minx = Math.min(minx, dist[row[i]][column[k]]);
            if (blank > 1) minx = 0;
            miny = Integer.MAX_VALUE;
            blank = 0;
            for (k=1; k<=elms; k++)
              if (dist[row[k]][column[j]] == 0)
                blank++;
              else
                miny = Math.min(miny, dist[row[k]][column[j]]);
            if (blank > 1) miny = 0;
            if ((minx + miny) > diff) {
              diff = minx + miny;
              candr = i;
              candc = j;
            }
          }
      thresh = weight + diff;
      front[row[candr]] = column[candc];
      back[column[candc]] = row[candr];
```

```
        tail = column[candc];
        while (front[tail] != 0)
          tail = front[tail];
        head = row[candr];
        while (back[head] != 0)
          head = back[head];
        stretch = dist[tail][head];
        dist[tail][head] = Integer.MAX_VALUE;
        for (i=1; i<=candr-1; i++)
          rowvec[i] = row[i];
        for (i=candr; i<=elms-1; i++)
          rowvec[i] = row[i+1];
        for (i=1; i<=candc-1; i++)
          colvec[i] = column[i];
        for (i=candc; i<=elms-1; i++)
          colvec[i] = column[i+1];
        tspsearch(nodes,edges+1,weight,dist,rowvec,colvec,cursol,front,back);
        dist[tail][head] = stretch;
        back[column[candc]] = 0;
        front[row[candr]] = 0;
        if (thresh < dist[0][0]) {
          dist[row[candr]][column[candc]] = Integer.MAX_VALUE;
          tspsearch(nodes,edges,weight,dist,row,column,cursol,front,back);
          dist[row[candr]][column[candc]] = 0;
        }
      }
    }
    for (i=1; i<=elms; i++)
      for (j=1; j<=elms; j++)
        dist[row[i]][column[j]] += (cutx[i] + cuty[j]);
  }
```

Example:

Find the minimum cost cycle for the following distance matrix with 5 nodes:

$$\begin{pmatrix} 999 & 27 & 21 & 90 & 45 \\ 12 & 999 & 69 & 11 & 73 \\ 50 & 14 & 999 & 55 & 26 \\ 55 & 66 & 71 & 999 & 42 \\ 96 & 34 & 84 & 52 & 999 \end{pmatrix}$$

```
package GraphAlgorithms;

public class Test_tsp extends Object {

  public static void main(String args[]) {
    int n = 5;
    int dist[][] = {{0,    0,   0,    0,    0,    0},
                    {0, 999,  27,   21,   90,   45},
                    {0,  12, 999,   69,   11,   73},
                    {0,  50,  14,  999,   55,   26},
                    {0,  55,  66,   71,  999,   42},
                    {0,  96,  34,   84,   52,  999}};
    int sol[] = new int[n + 1];
    GraphAlgo.travelingSalesmanProblem(n, dist, sol);
    System.out.print("Minimum cost cycle: ");
    for (int i=1; i<=n; i++)
      System.out.print("    " + sol[i]);
    System.out.println("\n\nTotal cost = " + dist[0][0]);
  }
}
```

Output:

```
Minimum cost cycle:    1   3   5   2   4

Total cost = 147
```

4. Planarity Testing

A graph is *planar* if it can be drawn on a plane without crossing any edges. The basic strategy in testing whether a graph G is planar involves finding a cycle C in G. Then the graph G-C is decomposed into edge-disjoint paths. The paths are added to cycle C one at a time while keeping the embedding planar. If the embedding is successful then G is planar, otherwise G is nonplanar.

Let n be the number of nodes and m be the number of edges in an undirected graph. The following planar testing algorithm [HT74] takes $O(n)$ operations.

Step 1. If $m > 3n-6$, then stop, the graph is nonplanar.

Step 2. Obtain a digraph D by a depth-first search on G so that the edges of G are divided into tree edges and backward edges. During the depth-first search compute the low point functions L1 and L2, where

L1(u) is the lowest node reachable from node u by a sequence of tree edges followed by at most one backward edge, and

L2(u) is the second lowest node reachable from node u in this way.

Step 3. Reorder the adjacency lists of D using a radix sort.

Step 4. Use the low point functions computed from Step 2 and the ordering of edges from Step 3 to choose a particular adjacency structure so that the nodes of D are numbered in the order they are reached during any depth-first search for D without changing the adjacency structure.

Step 5. Prepare for the path addition process to be done in Step 6 by performing a second depth-first search to select edges in the reverse order to that given by the adjacency structure.

Step 6. Perform a third depth-first search to generate one cycle and a number of edge-disjoint paths. Each generated path is added to a planar embedding that contains the cycle and all the previously generated paths. Note that either any two paths may not constrain each other, or they may constrain each other to have the same embedding or the opposite embedding. These dependency relations among paths can be viewed as a dependency gragh H.

Step 7. The dependency gragh H is two-colorable if and only if the original graph G is planar. Color the graph H by using a depth-first search. If H is two-colorable then return the result that G is planar, otherwise return the result that G is nonplanar.

Procedure parameters:

boolean planarityTesting (n, m, *nodei*, *nodej*)

planarityTesting: boolean;

exit: returns true if the input graph is planar, false otherwise.

n: int;

entry: the number of nodes of the undirected graph, labeled from 1 to n.

m: int;

entry: the number of edges in the graph.

nodei, *nodej*: int[m+1];

entry: *nodei[i]* and *nodej[i]* are the end nodes of the i-th edge in the graph, i=1,2,...,m.

```
public static boolean planarityTesting(int n, int m, int nodei[], int nodej[])
{
  int i, j, k, n1, n2, m2, nm2, n2m, nmp2, m7n5, m22, m33, mtotal;
  int node1, node2, qnode, tnode, tnum, aux1, aux2, aux3, aux4;

  int level[] = new int[1];
  int initp[] = new int[1];
  int snode[] = new int[1];
  int pnum[] = new int[1];
  int snum[] = new int[1];
  int nexte[] = new int[1];
  int store1[] = new int[1];
  int store2[] = new int[1];
  int store3[] = new int[1];
  int store4[] = new int[1];
  int store5[] = new int[1];
  int mark[] = new int[n + 1];
  int trail[] = new int[n + 1];
  int descendant[] = new int[n + 1];
  int firstlow[] = new int[n + 1];
  int secondlow[] = new int[n + 1];
  int nodebegin[] = new int[n + n + 1];
  int wkpathfind5[] = new int[m + 1];
  int wkpathfind6[] = new int[m + 1];
  int stackarc[] = new int[m + m + 1];
  int stackcolor1[] = new int[m + m + 3];
  int stackcolor2[] = new int[m + m + 3];
  int stackcolor3[] = new int[m + m + 3];
  int stackcolor4[] = new int[m + m + 3];
  int wkpathfind1[] = new int[m + m + 3];
  int wkpathfind2[] = new int[m + m + 3];
  int wkpathfind3[] = new int[m + m + m + 4];
  int wkpathfind4[] = new int[m + m + m + 4];
  int first[] = new int[n + m + m + 1];
  int second[] = new int[n + m + m + 1];
  int sortn[] = new int[n + n + m + 1];
  int sortptr1[] = new int[n + n + m + 1];
  int sortptr2[] = new int[n + n + m + 1];
  int start[] = new int[m - n + 3];
  int finish[] = new int[m - n + 3];
  int paint[] = new int[m - n + 3];
  int nextarc[] = new int[7*m - 5*n + 3];
  int arctop[] = new int[7*m - 5*n + 3];
  boolean middle[] = new boolean[1];
  boolean fail[] = new boolean[1];
  boolean examin[] = new boolean[m - n + 3];
  boolean arctype[] = new boolean[7*m - 5*n + 3];

  // check for the necessary condition
```

```
if (m > 3*n-6)
  return false;
n2 = n + n;
m2 = m + m;
nm2 = n + m + m;
n2m = n + n + m;
m22 = m + m + 2;
m33 = m + m + m + 3;
nmp2 = m - n + 2;
m7n5 = 7 * m - 5 * n + 2;
// set up graph representation
for (i=1; i<=n; i++)
  second[i] = 0;
mtotal = n;
for (i=1; i<=m; i++) {
  node1 = nodei[i];
  node2 = nodej[i];
  mtotal++;
  second[mtotal] = second[node1];
  second[node1] = mtotal;
  first[mtotal] = node2;
  mtotal++;
  second[mtotal] = second[node2];
  second[node2] = mtotal;
  first[mtotal] = node1;
}
// initial depth-first search, compute low point functions
for (i=1; i<=n; i++) {
  mark[i] = 0;
  firstlow[i] = n + 1;
  secondlow[i] = n + 1;
}
snum[0] = 1;
store1[0]  = 0;
mark[1] = 1;
wkpathfind5[1] = 1;
wkpathfind6[1] = 0;
level[0]  = 1;
middle[0] = false;
do {
  planarityDFS1(n,m,m2,nm2,level,middle,snum,store1,mark,firstlow,
                secondlow,wkpathfind5,wkpathfind6,stackarc,first,second);
} while (level[0] > 1);
for (i=1; i<=n; i++)
  if (secondlow[i] >= mark[i])  secondlow[i] = firstlow[i];
// radix sort
mtotal = n2;
k = n2;
for (i=1; i<=n2; i++)
  sortn[i] = 0;
```

```
for (i=2; i<=m2; i+=2) {
  k++;
  sortptr1[k] = stackarc[i-1];
  tnode = stackarc[i];
  sortptr2[k] = tnode;
  if (mark[tnode] < mark[sortptr1[k]]) {
    j = 2 * mark[tnode] - 1;
    sortn[k] = sortn[j];
    sortn[j] = k;
  }
  else {
    if (secondlow[tnode] >= mark[sortptr1[k]]) {
      j = 2 * firstlow[tnode] - 1;
      sortn[k] = sortn[j];
      sortn[j] = k;
    }
    else {
      j = 2 * firstlow[tnode];
      sortn[k] = sortn[j];
      sortn[j] = k;
    }
  }
}
for (i=1; i<=n2; i++) {
  j = sortn[i];
  while (j != 0) {
    node1 = sortptr1[j];
    node2 = sortptr2[j];
    mtotal++;
    second[mtotal] = second[node1];
    second[node1] = mtotal;
    first[mtotal] = node2;
    j = sortn[j];
  }
}
// second depth-first search
for (i=2; i<=n; i++)
  mark[i] = 0;
store1[0] = 0;
snum[0] = 1;
trail[1] = 1;
wkpathfind5[1] = 1;
start[1] = 0;
finish[1] = 0;
level[0] = 1;
middle[0] = false;
do {
  planarityDFS2(n,m,m2,nm2,level,middle,snum,store1,mark,
                wkpathfind5,stackarc,first,second);
} while (level[0] > 1);
```

```
mtotal = n;
for (i=1; i<=m; i++) {
  j = i + i;
  node1 = stackarc[j-1];
  node2 = stackarc[j];
  mtotal++;
  second[mtotal] = second[node1];
  second[node1] = mtotal;
  first[mtotal] = node2;
}
// path decomposition, construction of the dependency graph
store2[0]  = 0;
store3[0]  = 0;
store4[0]  = 0;
store5[0]  = 0;
initp[0] = 0;
pnum[0]  = 1;
wkpathfind1[1] = 0;
wkpathfind1[2] = 0;
wkpathfind2[1] = 0;
wkpathfind2[2] = 0;
wkpathfind3[1] = 0;
wkpathfind3[2] = n + 1;
wkpathfind3[3] = 0;
wkpathfind4[1] = 0;
wkpathfind4[2] = n + 1;
wkpathfind4[3] = 0;
for (i=1; i<=n2; i++)
  nodebegin[i] = 0;
nexte[0] = m - n + 1;
for (i=1; i<=m7n5; i++)
  nextarc[i] = 0;
snode[0] = n;
descendant[1] = n;
wkpathfind5[1] = 1;
level[0]  = 1;
middle[0] = false;
do {
  planarityDecompose(n,m,n2,m22,m33,nm2,nmp2,m7n5,level,middle,initp,
          snode,pnum,nexte,store2,store3,store4,store5,trail,descendant,
          nodebegin,wkpathfind5,start,finish,first,second,wkpathfind1,
          wkpathfind2,wkpathfind3,wkpathfind4,nextarc,arctop,arctype);
} while (level[0] > 1);
// perform two-coloring
pnum[0]--;
for (i=1; i<=nmp2; i++)
  paint[i] = 0;
j = pnum[0] + 1;
for (i=2; i<=j; i++)
  examin[i] = true;
```

```
tnum = 1;
while (tnum <= pnum[0]) {
  wkpathfind5[1] = tnum;
  paint[tnum] = 1;
  examin[tnum] = false;
  level[0] = 1;
  middle[0] = false;
  do {
    planarityTwoColoring(m,nmp2,m7n5,level,middle,fail,wkpathfind5,
                         paint,nextarc,arctop,examin,arctype);
    if (fail[0])
      return false;
  } while (level[0] > 1);
  while (!examin[tnum])
    tnum++;
}
aux1 = 0;
aux2 = 0;
aux3 = 0;
aux4 = 0;
stackcolor1[1] = 0;
stackcolor1[2] = 0;
stackcolor2[1] = 0;
stackcolor2[2] = 0;
stackcolor3[1] = 0;
stackcolor3[2] = 0;
stackcolor4[1] = 0;
stackcolor4[2] = 0;
for (i=1; i<=pnum[0]; i++) {
  qnode = start[i+1];
  tnode = finish[i+1];
  while (qnode <= stackcolor1[aux1+2])
    aux1 -= 2;
  while (qnode <= stackcolor2[aux2+2])
    aux2 -= 2;
  while (qnode <= stackcolor3[aux3+2])
    aux3 -= 2;
  while (qnode <= stackcolor4[aux4+2])
    aux4 -= 2;
  if (paint[i] == 1) {
    if (finish[trail[qnode]+1] != tnode) {
      if (tnode < stackcolor2[aux2+2])
        return false;
      if (tnode < stackcolor3[aux3+2])
        return false;
      aux3 += 2;
      stackcolor3[aux3 + 1] = i;
      stackcolor3[aux3 + 2] = tnode;
    }
    else {
```

```
        if ((tnode < stackcolor3[aux3+2]) &&
            (start[stackcolor3[aux3+1]+1] <= descendant[qnode]))
          return false;
        aux1 += 2;
        stackcolor1[aux1 + 1] = i;
        stackcolor1[aux1 + 2] = qnode;
      }
    }
    else {
      if (finish[trail[qnode]+1] != tnode) {
        if (tnode < stackcolor1[aux1+2])
          return false;
        if (tnode < stackcolor4[aux4+2])
          return false;
        aux4 += 2;
        stackcolor4[aux4 + 1] = i;
        stackcolor4[aux4 + 2] = tnode;
      }
      else {
        if ((tnode < stackcolor4[aux4+2]) &&
            (start[stackcolor4[aux4+1]+1] <= descendant[qnode]))
          return false;
        aux2 += 2;
        stackcolor2[aux2 + 1] = i;
        stackcolor2[aux2 + 2] = qnode;
      }
    }
  }
  return true;
}


static private void planarityDFS1(int n, int m, int m2, int nm2, int level[],
       boolean middle[], int snum[], int store1[], int mark[],
       int firstlow[], int secondlow[], int wkpathfind5[], int wkpathfind6[],
       int stackarc[], int first[], int second[])
{
  /* this method is used internally by planarityTesting */

  int pnode=0, qnode=0, tnode=0, tmp1, tmp2;
  boolean skip;

  skip = false;
  if (middle[0]) skip = true;
  if (!skip) {
    qnode = wkpathfind5[level[0]];
    pnode = wkpathfind6[level[0]];
  }
  while (second[qnode] > 0 || skip) {
    if (!skip) {
```

```
    tnode = first[second[qnode]];
    second[qnode] = second[second[qnode]];
  }
  if (((mark[tnode] < mark[qnode]) && (tnode != pnode)) || skip) {
    if (!skip) {
      store1[0]+= 2;
      stackarc[store1[0]-1] = qnode;
      stackarc[store1[0]]   = tnode;
    }
    if ((mark[tnode] == 0) || skip) {
      if (!skip) {
        snum[0]++;
        mark[tnode] = snum[0];
        level[0]++;
        wkpathfind5[level[0]] = tnode;
        wkpathfind6[level[0]] = qnode;
        middle[0] = false;
        return;
      }
      skip = false;
      tnode = wkpathfind5[level[0]];
      qnode = wkpathfind6[level[0]];
      level[0]--;
      pnode = wkpathfind6[level[0]];
      if (firstlow[tnode] < firstlow[qnode]) {
        tmp1 = secondlow[tnode];
        tmp2 = firstlow[qnode];
        secondlow[qnode] = (tmp1 < tmp2 ? tmp1 : tmp2);
        firstlow[qnode] = firstlow[tnode];
      }
      else {
        if (firstlow[tnode] == firstlow[qnode]) {
          tmp1 = secondlow[tnode];
          tmp2 = secondlow[qnode];
          secondlow[qnode] = (tmp1 < tmp2 ? tmp1 : tmp2);
        }
        else {
          tmp1 = firstlow[tnode];
          tmp2 = secondlow[qnode];
          secondlow[qnode] = (tmp1 < tmp2 ? tmp1 : tmp2);
        }
      }
    }
    else {
      if (mark[tnode] < firstlow[qnode]) {
        secondlow[qnode] = firstlow[qnode];
        firstlow[qnode] = mark[tnode];
      }
      else {
        if (mark[tnode] > firstlow[qnode]) {
```

```
            tmp1 = mark[tnode];
            tmp2 = secondlow[qnode];
            secondlow[qnode] = (tmp1 < tmp2 ? tmp1 : tmp2);
          }
        }
      }
    }
  }
  middle[0] = true;
}


static private void planarityDFS2(int n, int m, int m2, int nm2, int level[],
       boolean middle[], int snum[], int store1[], int mark[],
       int wkpathfind5[], int stackarc[], int first[], int second[])
{
  /* this method is used internally by planarityTesting */

  int qnode, tnode;

  if (middle[0]) {
    tnode = wkpathfind5[level[0]];
    level[0]--;
    qnode = wkpathfind5[level[0]];
    store1[0] += 2;
    stackarc[store1[0]-1] = mark[qnode];
    stackarc[store1[0]] = mark[tnode];
  }
  else
    qnode = wkpathfind5[level[0]];
  while (second[qnode] > 0) {
    tnode = first[second[qnode]];
    second[qnode] = second[second[qnode]];
    if (mark[tnode] == 0) {
      snum[0]++;
      mark[tnode] = snum[0];
      level[0]++;
      wkpathfind5[level[0]] = tnode;
      middle[0] = false;
      return;
    }
    store1[0] += 2;
    stackarc[store1[0]-1] = mark[qnode];
    stackarc[store1[0]] = mark[tnode];
  }
  middle[0] = true;
}
```

```
static private void planarityDecompose(int n, int m, int n2, int m22, int m33,
       int nm2, int nmp2, int m7n5, int level[], boolean middle[],
       int initp[], int snode[], int pnum[], int nexte[], int store2[],
       int store3[], int store4[], int store5[],  int trail[],
       int descendant[], int nodebegin[], int wkpathfind5[], int start[],
       int finish[], int first[], int second[], int wkpathfind1[],
       int wkpathfind2[], int wkpathfind3[], int wkpathfind4[], int nextarc[],
       int arctop[], boolean arctype[])
{
  /* this method is used internally by planarityTesting */

  int node1, node2, qnode=0, qnode2, tnode=0, tnode2;
  boolean ind, skip;

  skip = false;
  if (middle[0]) skip = true;
  if (!skip) qnode = wkpathfind5[level[0]];
  while ((second[qnode] != 0) || skip) {
    if (!skip) {
      tnode = first[second[qnode]];
      second[qnode] = second[second[qnode]];
      if (initp[0] == 0) initp[0] = qnode;
    }
    if ((tnode > qnode) || skip) {
      if (!skip) {
        descendant[tnode] = snode[0];
        trail[tnode] = pnum[0];
        level[0]++;
        wkpathfind5[level[0]] = tnode;
        middle[0] = false;
        return;
      }
      skip = false;
      tnode = wkpathfind5[level[0]];
      level[0]--;
      qnode = wkpathfind5[level[0]];
      snode[0] = tnode - 1;
      initp[0] = 0;
      while (qnode <= wkpathfind2[store3[0] + 2])
        store3[0] -= 2;
      while (qnode <= wkpathfind1[store2[0] + 2])
        store2[0] -= 2;
      while (qnode <= wkpathfind3[store4[0] + 3])
        store4[0] -= 3;
      while (qnode <= wkpathfind4[store5[0] + 3])
         store5[0] -= 3;
      ind = false;
      qnode2 = qnode + qnode;
      while ((nodebegin[qnode2 - 1] > wkpathfind3[store4[0] + 2])  &&
             (qnode < wkpathfind3[store4[0] + 2]) &&
```

```
          (nodebegin[qnode2] < wkpathfind3[store4[0] + 1])) {
      ind = true;
      node1 = nodebegin[qnode2];
      node2 = wkpathfind3[store4[0] + 1];
      nexte[0]++;
      nextarc[nexte[0]] = nextarc[node1];
      nextarc[node1] = nexte[0];
      arctop[nexte[0]] = node2;
      node1 = wkpathfind3[store4[0] + 1];
      node2 = nodebegin[qnode2];
      nexte[0]++;
      nextarc[nexte[0]] = nextarc[node1];
      nextarc[node1] = nexte[0];
      arctop[nexte[0]] = node2;
      arctype[nexte[0] - 1] = false;
      arctype[nexte[0]] = false;
      store4[0] -= 3;
    }
    if (ind) store4[0] += 3;
    nodebegin[qnode2 - 1] = 0;
    nodebegin[qnode2] = 0;
  }
  else {
    start[pnum[0] + 1] = initp[0];
    finish[pnum[0] + 1] = tnode;
    ind = false;
    if (wkpathfind1[store2[0]+2] != 0) {
      store3[0] += 2;
      wkpathfind2[store3[0]+1] = wkpathfind1[store2[0]+1];
      wkpathfind2[store3[0]+2] = wkpathfind1[store2[0]+2];
    }
    if (finish[wkpathfind1[store2[0]+1] + 1] != tnode) {
      while (tnode < wkpathfind2[store3[0]+2]) {
        node1 = pnum[0];
        node2 = wkpathfind2[store3[0] + 1];
        nexte[0]++;
        nextarc[nexte[0]] = nextarc[node1];
        nextarc[node1] = nexte[0];
        arctop[nexte[0]] = node2;
        node1 = wkpathfind2[store3[0] + 1];
        node2 = pnum[0];
        nexte[0]++;
        nextarc[nexte[0]] = nextarc[node1];
        nextarc[node1] = nexte[0];
        arctop[nexte[0]] = node2;
        arctype[nexte[0] - 1] = true;
        arctype[nexte[0]] = true;
        ind  = true;
        store3[0] -= 2;
      }
```

```
        if (ind) store3[0] += 2;
        ind = false;
        while ((tnode < wkpathfind3[store4[0]+3]) &&
               (initp[0] < wkpathfind3[store4[0]+2])) {
          node1 = pnum[0];
          node2 = wkpathfind3[store4[0] + 1];
          nexte[0]++;
          nextarc[nexte[0]] = nextarc[node1];
          nextarc[node1] = nexte[0];
          arctop[nexte[0]] = node2;
          node1 = wkpathfind3[store4[0] + 1];
          node2 = pnum[0];
          nexte[0]++;
          nextarc[nexte[0]] = nextarc[node1];
          nextarc[node1] = nexte[0];
          arctop[nexte[0]] = node2;
          arctype[nexte[0] - 1] = false;
          arctype[nexte[0]] = false;
          store4[0] -= 3;
        }
        while ((tnode < wkpathfind4[store5[0]+3])  &&
               (initp[0] < wkpathfind4[store5[0]+2]))
          store5[0] -= 3;
        tnode2 = tnode + tnode;
        if (initp[0] > nodebegin[tnode2-1]) {
          nodebegin[tnode2-1] = initp[0];
          nodebegin[tnode2] = pnum[0];
        }
        store4[0] += 3;
        wkpathfind3[store4[0]+1] = pnum[0];
        wkpathfind3[store4[0]+2] = initp[0];
        wkpathfind3[store4[0]+3] = tnode;
        store5[0] += 3;
        wkpathfind4[store5[0]+1] = pnum[0];
        wkpathfind4[store5[0]+2] = initp[0];
        wkpathfind4[store5[0]+3] = tnode;
      }
      else {
        while ((tnode < wkpathfind4[store5[0]+3])  &&
               (initp[0] < wkpathfind4[store5[0]+2])  &&
               (wkpathfind4[store5[0]+2] <= descendant[initp[0]])) {
          ind = true;
          node1 = pnum[0];
          node2 = wkpathfind4[store5[0] + 1];
          nexte[0]++;
          nextarc[nexte[0]] = nextarc[node1];
          nextarc[node1] = nexte[0];
          arctop[nexte[0]] = node2;
          node1 = wkpathfind4[store5[0] + 1];
          node2 = pnum[0];
```

```
          nexte[0]++;
          nextarc[nexte[0]] = nextarc[node1];
          nextarc[node1] = nexte[0];
          arctop[nexte[0]] = node2;
          arctype[nexte[0] - 1] = false;
          arctype[nexte[0]] = false;
          store5[0] -= 3;
        }
        if (ind) store5[0] += 3;
      }
      if (qnode != initp[0]) {
        store2[0] += 2;
        wkpathfind1[store2[0]+1] = pnum[0];
        wkpathfind1[store2[0]+2] = initp[0];
      }
      pnum[0]++;
      initp[0] = 0;
    }
  }
  middle[0] = true;
}


static private void planarityTwoColoring(int m, int nmp2, int m7n5, int level[],
               boolean middle[], boolean fail[], int wkpathfind5[], int paint[],
               int nextarc[], int arctop[], boolean examin[], boolean arctype[])
{
  /* this method is used internally by planarityTesting */

  int link, qnode, tnode;
  boolean dum1, dum2;

  fail[0] = false;
  if (middle[0]) {
    level[0]--;
    qnode = wkpathfind5[level[0]];
  }
  else
    qnode = wkpathfind5[level[0]];
  while (nextarc[qnode] != 0) {
    link  = nextarc[qnode];
    tnode = arctop[link];
    nextarc[qnode] = nextarc[link];
    if (paint[tnode] == 0)
      paint[tnode] = (arctype[link] ? paint[qnode] : 3 - paint[qnode]);
    else {
      dum1 = (paint[tnode] == paint[qnode]);
      dum2 = !arctype[link];
      if ((dum1 && dum2) || (!dum1 && !dum2)) {
        fail[0] = true;
```

```
        return;
      }
    }
    if (examin[tnode]) {
      examin[tnode] = false;
      level[0]++;
      wkpathfind5[level[0]] = tnode;
      middle[0] = false;
      return;
    }
  }
  middle[0] = true;
}
```

Example:

Check if the following graph is planar.

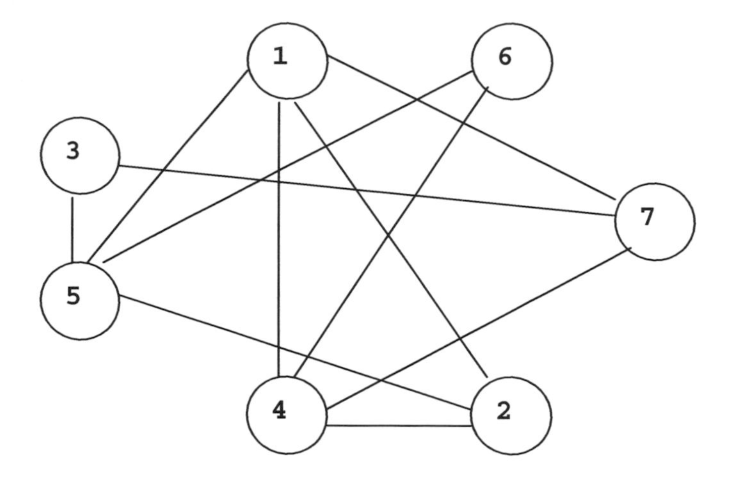


```
package GraphAlgorithms;

public class Test_planarity extends Object {

  public static void main(String args[]) {

    int n = 7;
    int m = 11;
    int nodei[] = {0, 1, 5, 2, 7, 1, 3, 5, 7, 4, 1, 1};
    int nodej[] = {0, 5, 2, 4, 4, 7, 5, 6, 3, 6, 4, 2};

    if (GraphAlgo.planarityTesting(n, m, nodei, nodej))
      System.out.println("The input graph is planar.");
    else
      System.out.println("The input graph is nonplanar.");
  }
}
```

Output:

```
The input graph is planar.
```

5. Graph Isomorphism Testing

Two graphs are said to be *isomorphic* if it is possible to relabel the nodes of one graph so that the two graphs are identical. The following procedure is from the FORTRAN90 library CODEPACK that tests whether two given undirected simple graphs are isomorphic [B00]. Two matrix indices of a matrix can be compared by using the following definition:

$$(u1, v1) < (u2, v2)$$

if either

$$\max(u1, v1) < \max(u2, v2)$$

or

$$\max(u1, v1) = \max(u2, v2) \text{ and } u1 < u2$$

Under this particular index order of the entries of a matrix, matrix P is less than or equal to matrix Q if either every entry of the P and Q is equal, or $P_{i,j} < Q_{i,j}$ in the first matrix entry (i,j) in which the matrices differ.

Define the graph code to the maximum matrix of all the adjacency matrices of a graph. The graph isomorphism is done by computing the codes of the two graphs. A backtracking process is used to examine all the possible permutations of the nodes of the graph. Start with the identity permutation of the nodes, all permutations in which the first node is swapped with the second. In some cases, if a portion of the adjacency matrix is enough to determine that the matrix would be smaller than the current best solution, then the current search can be shortened by moving on to the next candidate in the enumeration.

Procedure parameters:

int graphIsomorphism (*n*, *adj1*, *adj2*, *map1*, *map2*)

graphIsomorphism: int;
- exit: the method returns the following code:
  - 0: the two graphs are isomorphic
  - 1: nonisomorphic, different number of edges
  - 2: nonisomorphic, different node degree sequence
  - 3: nonisomorphic, different graph code
  - 4: nonisomorphic, graph code computation error

n: int;
- entry: the number of nodes in each graph, labeled from 1 to *n*.

adj1: int[*n*+1][*n*+1];
- entry: the adjacency matrix of the first graph with zero-one entries. Only input values of *adj1[i][j]*, i=1,2,...,*n*, j=1,2,...,*n*, are needed. The values for *adj1[0][j]* and *adj1[i][0]* are not required.

adj2: int[*n*+1][*n*+1];
- entry: the adjacency matrix of the second graph with zero-one entries. Only input values of *adj2[i][j]*, i=1,2,...,*n*, j=1,2,...,*n*, are needed. The values for *adj2[0][j]* and *adj2[i][0]* are not required.

map1, map2: int[*n*+1];
- exit: if the two graphs are isomorphic (this procedure returns zero), these two arrays return the relabeling of the nodes of the two graphs that realize the isomorphism. That is, the two graphs can

be shown to be identifcal by relabeling node *map1*[i] of the first graph to node *map2*[i], or by relabeling node *map2*[i] of the second graph to node *map1*[i], for i = 1, 2,...,*n*.

```
public static int graphIsomorphism(int n, int adj1[][], int adj2[][],
                                   int map1[], int map2[])
{
  int i,j,k,edges1,edges2;
  int label1[][] = new int[n + 1][n + 1];
  int label2[][] = new int[n + 1][n + 1];
  int degree1[] = new int[n + 1];
  int degree2[] = new int[n + 1];

  // validate the number of edges
  edges1 = 0;
  for (i=1; i<=n; i++)
    for (j=1; j<=n; j++)
      edges1 = (i == j) ? edges1 + 2 * adj1[i][j] : edges1 + adj1[i][j];
  edges1 /= 2;
  edges2 = 0;
  for (i=1; i<=n; i++)
    for (j=1; j<=n; j++)
      edges2 = (i == j) ? edges2 + 2 * adj2[i][j] : edges2 + adj2[i][j];
  edges2 /= 2;
  if (edges1 != edges2 ) return 1;
  // validate the degree sequences
  // node degrees of the first graph are ordered in decreasing order
  for (i=1; i<=n; i++)
    degree1[i] = 0;
  for (i=1; i<=n; i++)
    for (j=1; j<=n; j++)
      degree1[i] += adj1[i][j];
  // sort "degree1" in descending order
  GraphAlgo.heapsort(n, degree1, false);
  // node degree of the second graph are ordered in decreasing order
  for (i=1; i<=n; i++)
    degree2[i] = 0;
  for (i=1; i<=n; i++)
    for (j=1; j<=n; j++)
      degree2[i] += adj2[i][j];
  // sort "degree2" in descending order
  GraphAlgo.heapsort(n, degree2, false);
  // compare the degree sequence of the two graphs
  k = 1;
  while (k <= n) {
    if (degree1[k] < degree2[k]) return 2;
    k++;
  }
  // compute the code of the first graph
  if (!isomorphicCode(adj1, n, label1, map1)) return 4;
```

```
  // compute the code of the second graph
  if (!isomorphicCode(adj2, n, label2, map2)) return 4;
  // compare the codes of the two graphs
  for (j=2; j<=n; j++)
    for (i=1; i<=j-1; i++)
      if (label1[i][j] != label2[i][j]) return 3;
  return 0;
}


static private boolean isomorphicCode(int adj[][], int n,
                                      int label[][], int map[])
{
  /* this method is used internally by graphIsomorphism */

  int i,j,k,auxsize,flag,p,q,r,s,t,v;
  int ctr1[] = new int[1];
  int ctr2[] = new int[1];
  int ctr3[] = new int[1];
  int ctr4[] = new int[1];
  int savelabel[][] = new int[n+1][n+1];
  int savemap[] = new int[n+1];
  int aux1[] = new int[n+1];
  int aux2[] = new int[n+1];
  int aux3[] = new int[n+1];
  int aux4[] = new int[4 * n+1];
  boolean found;

  aux4[1] = 0;
  ctr3[0] = 0;
  ctr4[0] = 0;
  auxsize = 4 * n;
  // initial identity ordering
  for (i=1; i<=n; i++)
    map[i] = i;
  // compute the code
  for (i=1; i<=n; i++) {
    if (i <= n) {
      p = map[i];
      if (map[i] < 1 || n < map[i]) return false;
    }
    else
      p = 0;
    for (j=i+1; j<=n; j++) {
      if (j <= n) {
        q = map[j];
        if (map[j] < 1 || n < map[j]) return false;
      }
      else
        q = 0;
```

```
      if (p == 0 || q == 0)
        label[i][j] = label[j][i] = 0;
      else if ((p < q && adj[p][q] != 0 ) || (q < p && adj[q][p] != 0 ))
        label[i][j] = label[j][i] = 1;
      else
        label[i][j] = label[j][i] = 0;
    }
  }
  // save the current best ordering and code
  for (i=1; i<=n; i++) {
    for (j=1; j<=n; j++)
      savelabel[i][j] = label[i][j];
    savemap[i] = map[i];
  }
  // begin backtrack search
  // consider all possible orderings and their codes
  ctr1[0] = 0;
  while (true) {
    // construct the integer vector by backtracking
    if (ctr1[0] == 0) {
      // process the complete vector
      ctr2[0] = 1;
      ctr4[0] = 0;
      ctr1[0] = 2;
    }
    else {
      // examine the stack
      while (true) {
        if (0 < aux1[ctr2[0]]) {
          // take the first available one off the stack
          map[ctr2[0]] = aux4[ctr4[0]];
          ctr4[0]--;
          aux1[ctr2[0]]--;
          if (ctr2[0] != n) {
            ctr2[0]++;
            ctr1[0] = 2;
          }
          else
            ctr1[0] = 1;
          break;
        }
        else {
          // there are no candidates for position ctr2[0]
          ctr2[0]--;
          if (ctr2[0] <= 0) {
            // repeat the examination of the stack
            ctr1[0] = 3;
            break;
          }
        }
```

```
      }
    }
    // if the backtrack routine has returned a complete candidate
    // ordering, then compute the resulting code, and compare with
    // the current best and go back for the next backtrack search
    if (ctr1[0] == 1) {
      // compute the code
      for (i=1; i<=n; i++) {
        if (i <= n) {
          p = map[i];
          if (map[i] < 1 || n < map[i]) return false;
        }
        else
          p = 0;
        for (j=i+1; j<=n; j++) {
          if (j <= n) {
            q = map[j];
            if (map[j] < 1 || n < map[j]) return false;
          }
          else
            q = 0;
          if (p == 0 || q == 0)
            label[i][j] = label[j][i] = 0;
          else if ((p < q && adj[p][q] != 0 ) || (q < p && adj[q][p] != 0 ))
            label[i][j] = label[j][i] = 1;
          else
            label[i][j] = label[j][i] = 0;
        }
      }
      // compare savelabel and code
      flag = 0;
      for (j=2; j<=n; j++) {
        for (i=1; i<=j-1; i++)
          if (savelabel[i][j] < label[i][j]) {
            flag = - 1;
            break;
          }
          else if (label[i][j] < savelabel[i][j]) {
            flag = 1;
            break;
          }
        if (flag != 0) break;
      }
      ctr3[0]++;
      if (flag == -1) {
        for (i=1; i<=n; i++) {
          for (j=1; j<=n; j++)
            savelabel[i][j] = label[i][j];
          savemap[i] = map[i];
        }
```

```
      }
    }
    else if (ctr1[0] == 2) {
      // finds candidates for a maximal graph code ordering
      if (ctr2[0] < 1 || n < ctr2[0]) return false;
      aux1[ctr2[0]] = 0;
      found = false;
      if (1 < ctr2[0]) {
        // compute the graph code for this node ordering
        for (i=1; i<=n; i++) {
          if (i <= ctr2[0]-1) {
            p = map[i];
            if (map[i] < 1 || n < map[i]) return false;
          }
          else
            p = 0;
          for (j=i+1; j<=n; j++) {
            if (j <= ctr2[0]-1) {
              q = map[j];
              if (map[j] < 1 || n < map[j]) return false;
            }
            else
              q = 0;
            if (p == 0 || q == 0)
              label[i][j] = label[j][i] = 0;
            else if ((p < q && adj[p][q] != 0) || (q < p && adj[q][p] != 0))
              label[i][j] = label[j][i] = 1;
            else
              label[i][j] = label[j][i] = 0;
          }
        }
        // compares the two graph codes
        flag = 0;
        for (j=2; j<=ctr2[0]-1; j++) {
          for (i=1; i<=j-1; i++) {
            if (savelabel[i][j] < label[i][j]) {
              flag = - 1;
              break;
            }
            else if (label[i][j] < savelabel[i][j]) {
              flag = + 1;
              break;
            }
          }
          if (flag != 0) break;
        }
        ctr3[0]++;
        if (flag == 1) {
          aux1[ctr2[0]] = 0;
          found = true;
```

```
    }
  }
  if (!found) {
    // list of nodes that have not been used
    t = n + 1 - ctr2[0];
    // find the number of unused items in the permutation
    v = ctr2[0] - 1 + t;
    if ( ctr2[0] - 1 < 0 )
      return false;
    else if (ctr2[0]-1 == 0)
      for (i=1; i<=v; i++)
        aux2[i] = i;
    else if (t < 0)
      return false;
    else if (t == 0)
      {}
    else {
      k = 0;
      for (i=1; i<=v; i++) {
        r = 0;
        for (j=1; j<=ctr2[0]-1; j++)
          if (map[j] == i) {
            r = j;
            break;
          }
        if (r == 0) {
          k++;
          if (t < k) return false;
          aux2[k] = i;
        }
      }
    }
    aux1[ctr2[0]] = 0;
    for (i=1; i<=ctr2[0]-1; i++) {
      p = map[i];
      for (j=1; j<=t; j++) {
        q = aux2[j];
        if (adj[p][q] != 0 || adj[q][p] != 0) {
          aux1[ctr2[0]]++;
          ctr4[0]++;
          if (auxsize < ctr4[0]) return false;
          aux4[ctr4[0]] = q;
        }
      }
      if (0 < aux1[ctr2[0]]) {
        found = true;
        break;
      }
    }
    if (!found) {
```

```
        // no free nodes are connected to used nodes
        // take the free nodes with at least one neighbor
        s = 0;
        for (i=1; i<=t; i++) {
          p = aux2[i];
          aux3[i] = 0;
          for (j=1; j<=t; j++) {
            q = aux2[j];
            if (p != q)
              if (aux3[i] < adj[p][q]) aux3[i] = adj[p][q];
          }
          if (s < aux3[i]) s = aux3[i];
        }
        aux1[ctr2[0]] = 0;
        for (i=1; i<=t; i++)
          if (aux3[i] == s) {
            aux1[ctr2[0]]++;
            ctr4[0]++;
            if (auxsize < ctr4[0]) return false;
            aux4[ctr4[0]] = aux2[i];
          }
      }
    }
  }
  else
    // all possibilities have been examined
    break;
  }
  // set the best ordering and code
  for (i=1; i<=n; i++) {
    for (j=1; j<=n; j++)
      label[i][j] = savelabel[i][j];
    map[i] = savemap[i];
  }
  return true;
}


public static void heapsort(int n, int x[], boolean ascending)
{
  /* sort array elements x[1], x[2],..., x[n] in the order of
   *  increasing (ascending=true) or decreasing (ascending=false)
   */

  int elm,h,i,index,k,temp;

  if (n <= 1) return;
  // initially nodes n/2 to 1 are leaves in the heap
  for (i=n/2; i>=1; i--) {
    elm = x[i];
```

```
    index = i;
    while (true) {
      k = 2 * index;
      if (n < k) break;
      if (k + 1 <= n) {
        if (ascending) {
          if (x[k] < x[k+1]) k++;
        }
        else {
          if (x[k+1] < x[k]) k++;
        }
      }
      if (ascending) {
        if (x[k] <= elm) break;
      }
      else {
        if (elm <= x[k]) break;
      }
      x[index] = x[k];
      index = k;
    }
    x[index] = elm;
  }
  // swap x[1] with x[n]
  temp = x[1];
  x[1] = x[n];
  x[n] = temp;
  // repeat delete the root from the heap
  for (h=n-1; h>=2; h--) {
    // restore the heap structure of x[1] through x[h]
    for (i=h/2; i>=1; i--) {
      elm = x[i];
      index = i;
      while (true) {
        k = 2 * index;
        if (h < k) break;
        if (k + 1 <= h) {
          if (ascending) {
            if (x[k] < x[k+1]) k++;
          }
          else {
            if (x[k+1] < x[k]) k++;
          }
        }
        if (ascending) {
          if (x[k] <= elm) break;
        }
        else {
          if (elm <= x[k]) break;
        }
```

```
        x[index] = x[k];
        index = k;
      }
      x[index] = elm;
    }
    // swap x[1] and x[h]
    temp = x[1];
    x[1] = x[h];
    x[h] = temp;
  }
}
```

Example:

Test whether the following two graphs are isomorphic.

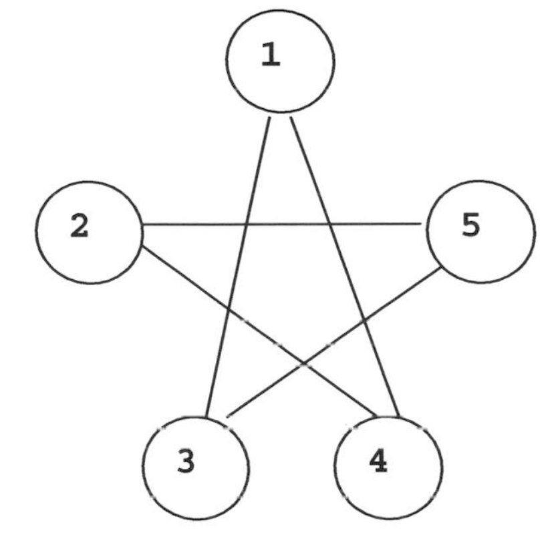

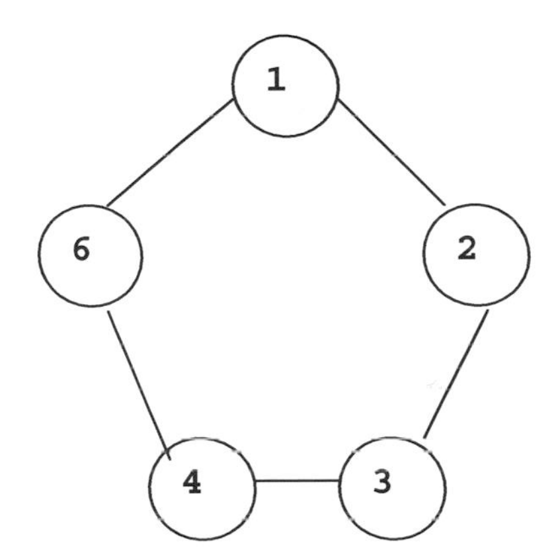

```
package GraphAlgorithms;

public class Test_graphIsomorphism extends Object {

  public static void main(String args[]) {

    int k;
    int n = 5;
    int map1[] = new int[n + 1];
    int map2[] = new int[n + 1];
    int adj1[][] = {{0, 0, 0, 0, 0, 0},
                    {0, 0, 0, 1, 1, 0},
                    {0, 0, 0, 0, 1, 1},
                    {0, 1, 0, 0, 0, 1},
                    {0, 1, 1, 0, 0, 0},
                    {0, 0, 1, 1, 0, 0}};
    int adj2[][] = {{0, 0, 0, 0, 0, 0},
                    {0, 0, 1, 0, 0, 1},
                    {0, 1, 0, 1, 0, 0},
                    {0, 0, 1, 0, 1, 0},
                    {0, 0, 0, 1, 0, 1},
                    {0, 1, 0, 0, 1, 0}};

    k = GraphAlgo.graphIsomorphism(n,adj1,adj2,map1,map2);
    if (k != 0)
      System.out.println("Input graphs are non-isomorphic, " +
                         "return code = " + k);
    else {
      System.out.print("Input graphs are isomorphic." +
                       "\n\n first graph node relabeling: ");
      for (k=1; k<=n; k++)
        System.out.print("  " + map1[k]);
      System.out.print("\n second graph node relabeling:");
      for (k=1; k<=n; k++)
        System.out.print("  " + map2[k]);
      System.out.println();
    }
  }
}
```

Output:

```
Input graphs are isomorphic.

 first graph node relabeling:   5  3  2  1  4
 second graph node relabeling:  5  4  1  3  2
```

6. Coloring

6.1 Node Coloring

A *node coloring* of an undirected graph G is an assignment of colors to nodes of G such that no adjacent nodes of G have the same color. A graph is *k-colorable* if there is a coloring of G using k colors. The chromatic number of G is the minimum k for which G is k-colorable. The procedure [B72] of node coloring is done by a simple implicit enumeration tree search method. Initially node 1 is assigned color 1, and the remaining nodes are colored sequentially so that node i is colored with the lowest numbered color which has not been used so far to color any nodes adjacent to node i.

Let p be the number of colors required by this feasible coloring. Try to generate a feasible coloring using $q < p$ colors. To accomplish this, all nodes colored with p must be recolored. Thus, a backtrack step can be taken up to node u, where node $u + 1$ is the lowest index assigned color p. Try to color node u with its smallest feasible alternative color greater than its current color. If there is no such alternative color which is smaller than p then backtrack to node $u - 1$. Otherwise, recolor node u and proceed forward, sequentially recoloring all nodes $u + 1$, $u + 2$, ..., with the smallest feasible color until either node n is colored or some node v is reached which requires color p. In the former case, an improved coloring using q colors has been found; in this case, backtrack and try to find a better coloring using less than q colors. In the latter case, backtrack from node v and proceed forward as before. The algorithm terminates when backtracking reaches node 1.

Procedure parameters:

void nodeColoring (*n, m, nodei, nodej, color*)

n: int;
entry: number of nodes of the undirected graph,
nodes of the graph are labeled from 1 to *n*.

m: int;
entry: number of edges of the undirected graph.

nodei, nodej: int[*m*+1];
entry: *nodei[p]* and *nodej[p]* are the end nodes of the p-th edge in the graph, p=1,2,...,*m*.

color: int[*n*+1];
exit: *colorl[0]* is the chromatic number of the graph.
color[i] is the color number assigned to node i, i=1,2,...,*n*.

```
public static void nodeColoring(int n, int m, int nodei[],
                                int nodej[], int color[])
{
  int i,j,k,loop,currentnode,newc,ncolor,paint,index,nodek,up,low;
  int degree[] = new int[n+1];
  int choices[] = new int[n+1];
  int maxcolornum[] = new int[n+1];
  int currentcolor[] = new int[n+1];
  int feasiblecolor[] = new int[n+1];
  int firstedges[] = new int[n+2];
  int endnode[] = new int[m+m+1];
  int availc[][] = new int[n+1][n+1];
  boolean more;

  // set up the forward star representation of the graph
  for (i=1; i<=n; i++)
    degree[i] = 0;
  k = 0;
  for (i=1; i<=n; i++) {
    firstedges[i] = k + 1;
    for (j=1; j<=m; j++)
      if (nodei[j] == i) {
        k++;
        endnode[k] = nodej[j];
        degree[i]++;
      }
      else {
        if (nodej[j] == i) {
          k++;
          endnode[k] = nodei[j];
          degree[i]++;
        }
      }
  }
  firstedges[n+1] = k + 1;
  for (i=1; i<=n; i++) {
    feasiblecolor[i] = degree[i] + 1;
    if (feasiblecolor[i] > i) feasiblecolor[i] = i;
    choices[i] = feasiblecolor[i];
    loop = feasiblecolor[i];
    for (j=1; j<=loop; j++)
      availc[i][j] = n;
    k = feasiblecolor[i] + 1;
    if (k <= n)
      for (j=k; j<=n; j++)
        availc[i][j] = 0;
  }
  currentnode = 1;
  // color currentnode
  newc = 1;
```

```
ncolor = n;
paint = 0;
more = true;
do {
  if (more) {
    index = choices[currentnode];
    if (index > paint + 1) index = paint + 1;
    while ((availc[currentnode][newc] < currentnode) && (newc <= index))
      newc++;
    // currentnode has the color 'newc'
    if (newc == index + 1)
      more = false;
    else {
      if (currentnode == n) {
        // a new coloring is found
        currentcolor[currentnode] = newc;
        for (i=1; i<=n; i++)
          color[i] = currentcolor[i];
        if (newc > paint) paint++;
        ncolor = paint;
        if (ncolor > 2) {
        // backtrack to the first node of color 'ncolor'
          index = 1;
          while (color[index] != ncolor)
            index++;
          j = n;
          while (j >= index) {
            currentnode--;
            newc = currentcolor[currentnode];
            paint = maxcolornum[currentnode];
            low = firstedges[currentnode];
            up = firstedges[currentnode + 1];
            if (up > low) {
              up--;
              for (k=low; k<=up; k++) {
                nodek = endnode[k];
                if (nodek > currentnode)
                  if (availc[nodek][newc] == currentnode) {
                    availc[nodek][newc] = n;
                    feasiblecolor[nodek]++;
                  }
              }
            }
            newc++;
            more = false;
            j--;
          }
          paint = ncolor - 1;
          for (i=1; i<=n; i++) {
            loop = choices[i];
```

```
          if (loop > paint) {
            k = paint + 1;
            for (j=k; j<=loop; j++)
              if (availc[i][j] == n) feasiblecolor[i]--;
            choices[i] = paint;
          }
        }
      }
    }
    else {
      // currentnode is less than n
      low = firstedges[currentnode];
      up = firstedges[currentnode + 1];
      if (up > low) {
        up--;
        k = low;
        while ((k <= up) && more) {
          nodek = endnode[k];
          if (nodek > currentnode)
            more = !((feasiblecolor[nodek] == 1) &&
                   (availc[nodek][newc] >= currentnode));
          k++;
        }
      }
      if (more) {
        currentcolor[currentnode] = newc;
        maxcolornum[currentnode] = paint;
        if (newc > paint) paint++;
        low = firstedges[currentnode];
        up = firstedges[currentnode + 1];
        if (up > low) {
          up--;
          for (k=low; k<=up; k++) {
            nodek = endnode[k];
            if (nodek > currentnode)
              if (availc[nodek][newc] >= currentnode) {
                availc[nodek][newc] = currentnode;
                feasiblecolor[nodek]--;
              }
          }
        }
        currentnode++;
        newc = 1;
      }
      else
        newc++;
    }
  }
}
else {
```

```
      more = true;
      if ((newc > choices[currentnode]) || (newc > paint + 1)) {
        currentnode--;
        newc = currentcolor[currentnode];
        paint = maxcolornum[currentnode];
        low = firstedges[currentnode];
        up = firstedges[currentnode + 1];
        if (up > low) {
          up--;
          for (k=low; k<=up; k++) {
            nodek = endnode[k];
            if (nodek > currentnode)
              if (availc[nodek][newc] == currentnode) {
                availc[nodek][newc] = n;
                feasiblecolor[nodek]++;
              }
          }
        }
        newc++;
        more = false;
      }
    }
  } while ((currentnode != 1) && (ncolor != 2));
  color[0] = ncolor;
}
```

Example:

Find the node coloring of the following graph.

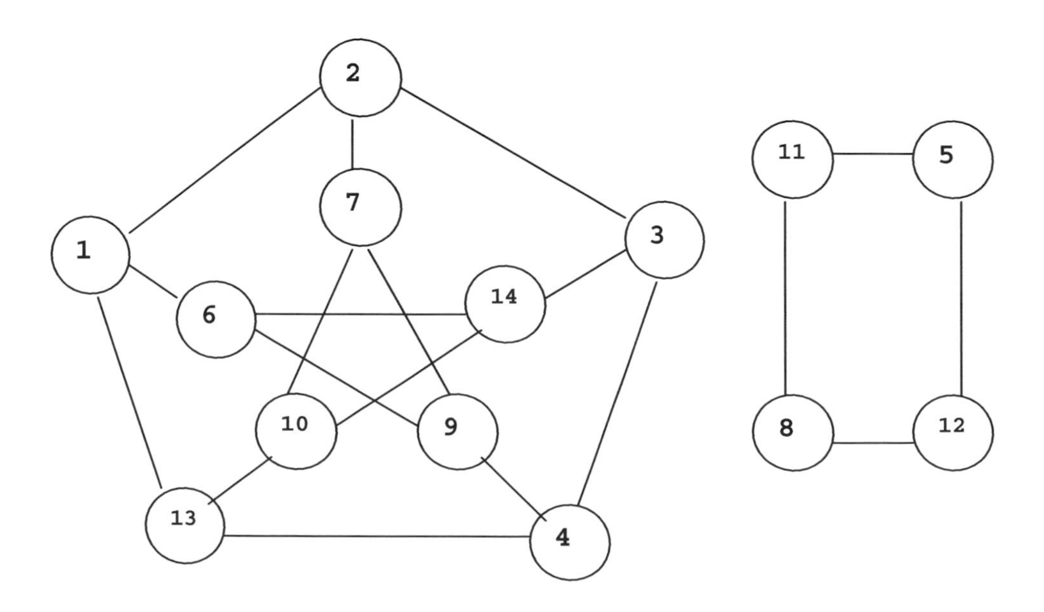

```
package GraphAlgorithms;

public class Test_nodeColoring extends Object {

  public static void main(String args[]) {

    int n=14, m=19;
    int color[] = new int[n+1];
    int nodei[] = {0,2,13,1,2,7,3, 3, 4,9,13,14,6,9, 7,14,11, 5,12, 5};
    int nodej[] = {0,1, 1,6,3,2,4,14,13,4,10, 6,9,7,10,10, 8,12, 8,11};

    GraphAlgo.nodeColoring(n,m,nodei,nodej,color);
    System.out.print("The chromatic number is " + color[0] + "\n\n  Node : ");
    for (int i=1; i<=n; i++)
      System.out.printf("%3d",i);
    System.out.print("\n  Color: ");
    for (int i=1; i<=n; i++)
      System.out.printf("%3d",color[i]);
    System.out.println();
  }
}
```

Output:

```
The chromatic number is 3

  Node :   1  2  3  4  5  6  7  8  9 10 11 12 13 14
  Color:   1  2  1  2  1  2  1  1  3  2  2  2  3  3
```

6.2 Chromatic Polynomial

Let f(p;G) be the number of colorings of a connected graph G in p or few colors. For any graph G with n nodes, f(p;G) is a polynomial in p. The *chromatic polynomial* f(p;G) can be expressed in three common forms:

1. $f(p;G) = a_n p^n - a_{n-1} p^{n-1} + a_{n-2} p^{n-2} - \cdots + (-1)^{n-1} a_1 p$
   in which the coefficients $a_i \geq 0$, i=1,2,...,n−1 alternate in sign, and $a_n = 1$.
2. $f(p;G) = \sum_{i=1}^{n} (-1)^{n-i} b_i p(p-1)^{i-1}$
   where $b_i \geq 0$, i=1,2,...,n.
3. $f(p;G) = \sum_{i=1}^{n} c_i p_{(i)}$
   where $p_{(i)} = p(p-1)(p-2)\ldots(p-i+1)$.

Consider two adjacent nodes u and v in a graph G. Let H be the graph obtained from G by deleting the edge (u,v) from G, and let F be the graph obtained from G by merging the two nodes u and v in G. Thus,

$$f(p;G) = f(p;H) - f(p;F).$$

The chromatic polynomial f(p;G) can be obtained by repeated application of this reduction formula, thereby expressing f(p;G) as a linear combination of chromatic polynomials of null graphs. The complete derivation can be represented by a binary tree in which left and right branches correspond to the deleting and merging steps, respectively. The running time of the algorithm [R68] is $O(m*2^m)$, where m is the number of edges in the input graph.

Procedure parameters:

void chromaticPolynomial (*n, m, nodei, nodej, cpoly1, cpoly2, cpoly3*)

n: int;
entry: number of nodes of the undirected graph, nodes of the graph are labeled from 1 to *n*.

m: int;
entry: number of edges of the undirected graph.

nodei, nodej: int[*m*+1];
entry: *nodei[p]* and *nodej[p]* are the end nodes of the p-th edge in the graph, p=1,2,...,*m*.

cpoly1: int[*n*+1];
exit: coefficients a_i for i=1,2,...,n, of the chromatic polynomial in the first form described above.

cpoly2: int[*n*+1];
exit: coefficients b_i for i=1,2,...,n, of the chromatic polynomial in the second form described above.

cpoly3: int[*n*+1];
exit: coefficients c_i for i=1,2,...,n, of the chromatic polynomial in the third form described above.

```
public static void chromaticPolynomial(int n, int m, int nodei[],
          int nodej[], int cpoly1[], int cpoly2[], int cpoly3[])
{
  int i,j,k,mm,nn,maxmn,ncomp,index,nodeu,nodev,nodew,nodex,incr;
  int isub2,jsub2,ivertex,jvertex,loop,top,ilast,jlast;
  int isub1=0,jsub1=0,ix=0,iy=0,nodey=0;
  int istack[] = new int[((n*(m+m-n+1))/2)+1];
  int jstack[] = new int[((n*(m+m-n+1))/2)+1];
  boolean visit,nonpos,skip;

  top = 0;
  for (i=1; i<=n; i++)
    cpoly2[i] = 0;
  mm = m;
  nn = n;
  // find a spanning tree
  while (true) {
```

```
maxmn = mm + 1;
if (nn > mm) maxmn = nn + 1;
for (i=1; i<=nn; i++)
  cpoly3[i] = -i;
for (i=1; i<=mm; i++) {
  j = nodei[i];
  nodei[i] = cpoly3[j];
  cpoly3[j] = - maxmn - i;
  j = nodej[i];
  nodej[i] = cpoly3[j];
  cpoly3[j] = - maxmn - maxmn - i;
}
ncomp = 0;
index = 0;
nodew = 0;
while (true) {
  nodew++;
  if (nodew > nn) break;
  nodev = cpoly3[nodew];
  if (nodev > 0) continue;
  ncomp++;
  cpoly3[nodew] = ncomp;
  if (nodev >= -nodew) continue;
  nodeu = nodew;
  nodex = -nodev;
  visit = true;
  isub2 = -nodev / maxmn;
  jsub2 = -nodev - isub2 * maxmn;
  while (true) {
    nodev = (isub2 == 1) ? nodei[jsub2] : nodej[jsub2];
    if (nodev > 0)
      if (nodev <= maxmn)
         if (cpoly3[nodev] >= 0) nodey = jsub2;
    if (nodev >= 0) {
      if (ix == 1) {
        nodex = Math.abs(nodei[iy]);
         nodei[iy] = nodeu;
      }
      else {
        nodex = Math.abs(nodej[iy]);
        nodej[iy] = nodeu;
      }
      ix = nodex / maxmn;
      iy = nodex - ix * maxmn;
      if (ix == 0) {
        skip = false;
        do {
          if (nodey != 0) {
            index++;
            cpoly3[nodeu] = nodei[nodey] + nodej[nodey] - nodeu;
```

```
            nodei[nodey] = -index;
            nodej[nodey] = nodeu;
          }
          nodeu = iy;
          if (iy <= 0) {
            skip = true;
            break;
          }
          nodex = -cpoly3[nodeu];
          ix = nodex / maxmn;
          iy = nodex - ix * maxmn;
        } while (ix == 0);
        if (skip) break;
      }
      isub2 = 3 - ix;
      jsub2 = iy;
      continue;
    }
    if (nodev < -maxmn) {
      if (isub2 == 1)
        nodei[jsub2] = -nodev;
      else
        nodej[jsub2] = -nodev;
      isub2 = -nodev / maxmn;
      jsub2 = -nodev - isub2 * maxmn;
    }
    else {
      nodev = - nodev;
      if (isub2 == 1)
        nodei[jsub2] = 0;
      else
        nodej[jsub2] = 0;
      if (visit) {
        isub1 = isub2;
        jsub1 = jsub2;
        nodey = 0;
        visit = false;
      }
      else {
        if (isub1 == 1)
          nodei[jsub1] = nodev;
        else
          nodej[jsub1] = nodev;
        isub1 = isub2;
        jsub1 = jsub2;
        if (ix == 1) {
          nodex = Math.abs(nodei[iy]);
          nodei[iy] = nodeu;
        }
        else {
```

```
            nodex = Math.abs(nodej[iy]);
            nodej[iy] = nodeu;
          }
        }
        ix = nodex / maxmn;
        iy = nodex - ix * maxmn;
        if (ix == 0) {
          skip = false;
          do {
            if (nodey != 0) {
              index++;
              cpoly3[nodeu] = nodei[nodey] + nodej[nodey] - nodeu;
              nodei[nodey] = -index;
              nodej[nodey] = nodeu;
            }
            nodeu = iy;
            if (iy <= 0) {
              skip = true;
              break;
            }
            nodex = -cpoly3[nodeu];
            ix = nodex / maxmn;
            iy = nodex - ix * maxmn;
          } while (ix == 0);
          if (skip) break;
        }
        isub2 - 3 - ix;
        jsub2 = iy;
      }
    }
  }
  for (i=1; i<=mm; i++) {
    while (true) {
      nodey = -nodei[i];
      if (nodey < 0) break;
      nodex = nodej[i];
      nodej[i] = nodej[nodey];
      nodej[nodey] = nodex;
      nodei[i] = nodei[nodey];
      nodei[nodey] = cpoly3[nodej[nodey]];
    }
  }
  for (i=1; i<=index; i++)
    cpoly3[nodej[i]] = cpoly3[nodei[i]];
  // if ncomp is not equal to 1, the graph is not connected
  if (ncomp != 1) break;
  if (mm < nn) {
    cpoly2[nn]++;
    if (top == 0) break;
    nn = istack[top];
```

```
      mm = jstack[top];
      top -= mm + 1;
      for (i=1; i<=mm; i++) {
        nodei[i] = istack[top + i];
        nodej[i] = jstack[top + i];
      }
      if (mm == nn)
        cpoly2[nn]++;
      else {
        top += mm;
        istack[top] = nn;
        jstack[top] = mm - 1;
      }
    }
    else {
      if (mm == nn)
        cpoly2[nn]++;
      else {
        for (i=1; i<=mm; i++) {
          top++;
          istack[top] = nodei[i];
          jstack[top] = nodej[i];
        }
        istack[top] = nn;
        jstack[top] = mm - 1;
      }
    }
    for (i=1; i<=n; i++)
      cpoly1[i] = 0;
    ivertex = (nodei[mm] < nodej[mm]) ? nodei[mm] : nodej[mm];
    jvertex = nodei[mm] + nodej[mm] - ivertex;
    loop = mm - 1;
    mm  = 0;
    for (i=1; i<=loop; i++) {
      ilast = nodei[i];
      if (ilast == jvertex) ilast = ivertex;
      if (ilast == nn) ilast = jvertex;
      jlast = nodej[i];
      if (jlast == jvertex) jlast = ivertex;
      if (jlast == nn) jlast = jvertex;
      if (ilast == ivertex) {
        if (cpoly1[jlast] != 0) continue;
        cpoly1[jlast] = 1;
      }
      if (jlast == ivertex) {
        if (cpoly1[ilast] != 0) continue;
        cpoly1[ilast] = 1;
      }
      mm++;
      nodei[mm] = ilast;
```

```
        nodej[mm] = jlast;
      }
      nn--;
    }
    for (i=1; i<=n; i++) {
      cpoly1[i] = cpoly2[i];
      cpoly3[i] = cpoly2[i] * (1 - 2 * ((n-i) - ((n-i)/2) * 2));
    }
    for (i=1; i<=n; i++) {
      jvertex = 0;
      for (j=i; j<=n; j++) {
        jvertex = cpoly1[n + i - j] + jvertex;
        cpoly1[n + i - j] = jvertex;
      }
    }
    incr = 0;
    for (i=1; i<=n; i++) {
      jvertex = 0;
      for (j=i; j<=n; j++) {
        jvertex = cpoly3[n + i - j] + incr * jvertex;
        cpoly3[n + i - j] = jvertex;
      }
      incr++;
    }
  }
```

Example:

Find the three forms of the chromatic polynomials of the following graph.

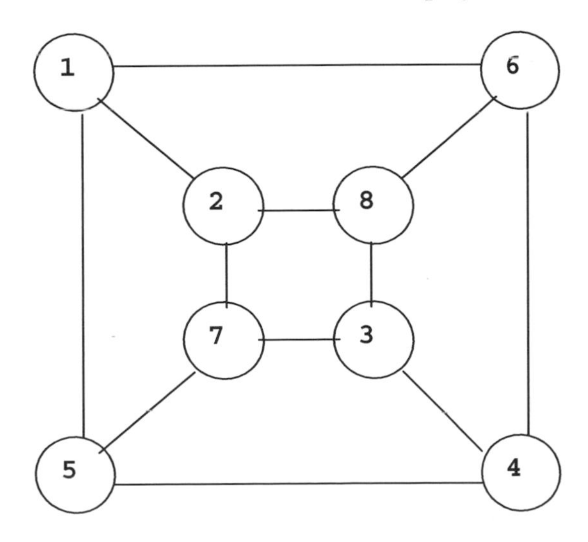

```
package GraphAlgorithms;

public class Test_chromaticPolynomial extends Object {

  public static void main(String args[]) {

    int n=8, m=12;
    int chpoly1[] = new int[n+1];
    int chpoly2[] = new int[n+1];
    int chpoly3[] = new int[n+1];
    int nodei[] = {0, 1, 1, 1, 2, 2, 3, 3, 3, 4, 4, 5, 6};
    int nodej[] = {0, 2, 5, 6, 7, 8, 4, 7, 8, 5, 6, 7, 8};

    GraphAlgo.chromaticPolynomial(n,m,nodei,nodej,chpoly1,chpoly2,chpoly3);
    System.out.print("Chromatic polynomial 1st form: ");
    for (int i=1; i<=n; i++)
      System.out.printf("%4d",chpoly1[i]);
    System.out.print("\nChromatic polynomial 2nd form: ");
    for (int i=1; i<=n; i++)
      System.out.printf("%4d",chpoly2[i]);
    System.out.print("\nChromatic polynomial 3rd form: ");
    for (int i=1; i<=n; i++)
      System.out.printf("%4d",chpoly3[i]);
    System.out.println();
  }
}
```

Output:

```
Chromatic polynomial 1st form:  133 423 572 441 214  66  12   1
Chromatic polynomial 2nd form:    0  11  32  40  29  15   5   1
Chromatic polynomial 3rd form:    0   1  18  92 146  80  16   1
```

7. Graph Matching

7.1 Maximum Cardinality Matching

For a given undirected graph G, a *matching* is a subset S of edges of G in which no two edges in S are adjacent in G. The *maximum cardinality matching* problem is to find a maching of maximum cardinality.

A node that is not matched is called an *exposed* node. An *alternating path* with respect to a given matching S is a path in which the edges appear alternately in S and not in S. An *augmenting path* is an alternating path which begins with an exposed node and ends with another exposed node. A fundamental theorem [B57] states that a matching S in G is maximum if and only if G has no augmenting path with respect to S.

The basic method for finding the maximum matching starts with an arbitrary matching Q. An augmenting path P with respect to Q is found. A new matching is constructed by taking those edges of P or Q that are not in both P and Q. The process is repeated and the matching is maximum when no augmenting path is found. The algorithm [E65, PC79] takes $O(n^3)$ operations for a graph of n nodes.

Procedure parameters:

void cardinalityMatching (*n, m, nodei, nodej, pair*)

n: int;
  entry: the number of nodes of the graph, labeled from 1 to *n*.

m: int;
  entry: the number of edges in the graph.

nodei, nodej: int[*m*+1];
  entry: *nodei[i]* and *nodej[i]* are the end nodes of the i-th edge in the graph, i=1,2,...,*m*. Note that the graph is not necessarily connected.

pair: int[*n*+1];
  exit: in the solution, node *i* is matched with *pair[i]*, i=1,2,...,*n*. If *pair[i]* = 0 then node *i* is unmatched. The number of unmatched nodes is given by *pair[0]*.

```
public static void cardinalityMatching(int n, int m, int nodei[],
                                       int nodej[], int pair[])
{
  int i, j, k, n1, istart, first, last, nodep, nodeq, nodeu, nodev, nodew;
  int neigh1, neigh2, unmatch;
  int fwdedge[] = new int[m + m + 1];
  int firstedge[] = new int[n + 2];
  int grandparent[] = new int[n + 1];
  int queue[] = new int[n + 1];
  boolean outree[] = new boolean[n + 1];
  boolean newnode, nopath;
```

```
// set up the forward star graph representation
n1 = n + 1;
k = 0;
for (i=1; i<=n; i++) {
  firstedge[i] = k + 1;
  for (j=1; j<=m; j++)
    if (nodei[j] == i) {
      k++;
      fwdedge[k] = nodej[j];
    }
    else {
      if (nodej[j] == i) {
        k++;
        fwdedge[k] = nodei[j];
      }
    }
}
firstedge[n1] = m + 1;

// all nodes are unmatched
unmatch = n;
for (i=1; i<=n; i++)
  pair[i] = 0;
for (i=1; i<=n; i++)
  if (pair[i] == 0) {
    j = firstedge[i];
    k = firstedge[i + 1] - 1;
    while ((pair[fwdedge[j]] != 0) && (j < k))
      j++;
    if (pair[fwdedge[j]] == 0) {
      // match a pair of nodes
      pair[fwdedge[j]] = i;
      pair[i] = fwdedge[j];
      unmatch -= 2;
    }
  }
for (istart=1; istart<=n; istart++)
  if ((unmatch >= 2) && (pair[istart] == 0)) {
    // 'istart' is not yet matched
    for (i=1; i<=n; i++)
      outree[i] = true;
    outree[istart] = false;
    // insert the root in the queue
    queue[1] = istart;
    first = 1;
    last = 1;
    nopath = true;
    do {
      nodep = queue[first];
      first = first + 1;
```

```
        nodeu = firstedge[nodep];
        nodew = firstedge[nodep + 1] - 1;
        while (nopath && (nodeu <= nodew)) {
          // examine the neighbor of 'nodep'
          if (outree[fwdedge[nodeu]]) {
            neigh2 = fwdedge[nodeu];
            nodeq = pair[neigh2];
            if (nodeq == 0) {
              // an augmentation path is found
              pair[neigh2] = nodep;
              do {
                neigh1 = pair[nodep];
                pair[nodep] = neigh2;
                if (neigh1 != 0) {
                  nodep = grandparent[nodep];
                  pair[neigh1] = nodep;
                  neigh2 = neigh1;
                }
              } while (neigh1 != 0);
              unmatch -= 2;
              nopath = false;
            }
            else {
              if (nodeq != nodep) {
                if (nodep == istart)
                   newnode = true;
                else {
                  nodev = grandparent[nodep];
                  while ((nodev != istart) && (nodev != neigh2))
                     nodev = grandparent[nodev];
                  newnode = (nodev == istart ? true : false);
                }
                if (newnode) {
                  // add a tree link
                  outree[neigh2] = false;
                  grandparent[nodeq] = nodep;
                  last++;
                  queue[last] = nodeq;
                }
              }
            }
          }
          nodeu++;
        }
      } while (nopath && (first <= last));
    }
  pair[0] = unmatch;
}
```

Example:

Find the maximum cardinality matching of the following graph.

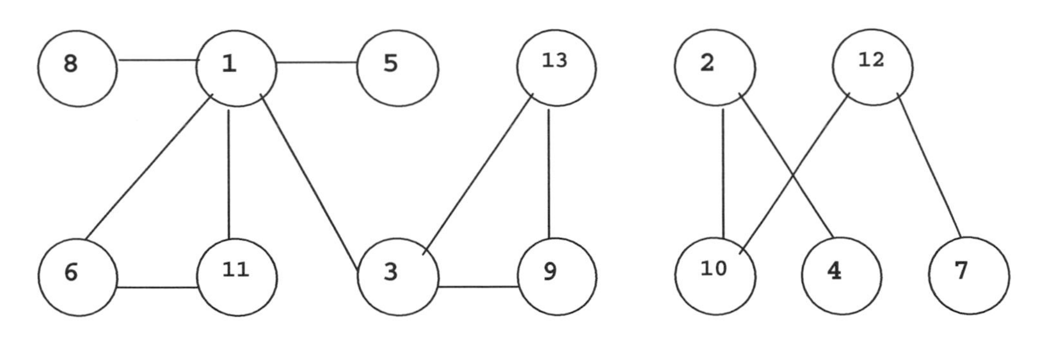


```
package GraphAlgorithms;

public class Test_cardinalityMatching extends Object {

  public static void main(String args[]) {

    int n = 13;
    int m = 13;
    int inode[] = {0, 4, 3, 1,12, 6, 8,13,10, 1, 9, 1,10,3};
    int jnode[] = {0, 2,13, 5, 7,11, 1, 9,12, 6, 3,11, 2,1};
    int pair[] = new int[n + 1];

    GraphAlgo.cardinalityMatching(n, m, inode, jnode, pair);
    System.out.println("Maximum matching:\n");
    for (int i=1; i<=n; i++)
      System.out.print("  " + pair[i]);
    System.out.println("\n\nNumber of unmatched nodes = " + pair[0]);
  }
}
```

Output:

```
Maximum matching:

  5  4  13  2  1  11  12  0  0  0  6  7  3

Number of unmatched nodes = 3
```

7.2 Minimum Sum Perfect Matching

For a given undirected graph G, a *matching* is a subset S of edges of G in which no two edges in S are adjacent in G. A matching is *perfect* if every vertex is incident to a matching edge. Consider a graph with n vertices, where n is even. Suppose the nonnegative edge weight for every pair of nodes i and j is given. The *minimum sum perfect matching* problem is to find a perfect matching in G such that the sum of the weights of the matching edges is minimum.

In a given matching of a graph G, a node that is not matched is called an *exposed* node. An *alternating path* with respect to a given matching S is a path in which the edges appear alternately in S and not in S. An *augmenting path* is an alternating path which begins with an exposed node and ends with another exposed node. The following procedure [BD80] solves the minimum sum perfect matching problem by iteratively augmenting along the shortest augmenting paths using a labeling and shrinking procedure [EJ73].

Procedure parameters:

void minSumMatching (*n, weight, sol*)

n: int;
entry: *n* is even, the number of nodes of the undirected graph labeled from 1 to *n*.

weight: double[*n*+1][*n*+1];
entry: the nonnegative edge weights of the undirected graph are stored in the upper triangular matrix *weight*. *weight[i][j]* is the weight of the edge between node i and node j, i<j. The values of the diagonal elements and the lower triangular entries of the matrix *weight* are not needed as input data.
exit: *weight[0][0]* is the total weight of the edges in the optimal perfect matching.

sol: int[*n*+1];
exit: in the optimal solution, node *i* is matched with node *sol[i]*, i=1,2,...,*n*.

```
public static void minSumMatching(int n, double weight[][], int sol[])
{
  int nn,i,j,head,min,max,sub,idxa,idxc;
  int kk1,kk3,kk6,mm1,mm2,mm3,mm4,mm5;
  int index=0,idxb=0,idxd=0,idxe=0,kk2=0,kk4=0,kk5=0;
  int aux1[] = new int[n+(n/2)+1];
  int aux2[] = new int[n+(n/2)+1];
  int aux3[] = new int[n+(n/2)+1];
  int aux4[] = new int[n+1];
  int aux5[] = new int[n+1];
  int aux6[] = new int[n+1];
  int aux7[] = new int[n+1];
```

```
int aux8[] = new int[n+1];
int aux9[] = new int[n+1];
double big,eps,cswk,cwk2,cst,cstlow,xcst,xwork,xwk2,xwk3,value;
double work1[] = new double[n+1];
double work2[] = new double[n+1];
double work3[] = new double[n+1];
double work4[] = new double[n+1];
double cost[] = new double[n*(n-1)/2 + 1];
boolean fin,skip;

// initialization
eps = 1.0e-5;
fin = false;
nn = 0;
for (j=2; j<=n; j++)
  for (i=1; i<j; i++) {
    nn++;
    cost[nn] = weight[i][j];
  }
big = 1.;
for (i=1; i<=n; i++)
  big += cost[i];
aux1[2] = 0;
for (i=3; i<=n; i++)
  aux1[i] = aux1[i-1] + i - 2;
head = n + 2;
for (i=1; i<=n; i++) {
  aux2[i] = i;
  aux3[i] = i;
  aux4[i] = 0;
  aux5[i] = i;
  aux6[i] = head;
  aux7[i] = head;
  aux8[i] = head;
  sol[i] = head;
  work1[i] = big;
  work2[i] = 0.;
  work3[i] = 0.;
  work4[i] = big;
}
// start procedure
for (i=1; i<=n; i++)
  if (sol[i] == head) {
    nn = 0;
    cwk2 = big;
    for (j=1; j<=n; j++) {
      min = i;
      max = j;
      if (i != j) {
        if (i > j) {
```

```
          max = i;
          min = j;
        }
        sub = aux1[max] + min;
        xcst = cost[sub];
        cswk = cost[sub] - work2[j];
        if (cswk <= cwk2) {
          if (cswk == cwk2) {
            if (nn == 0)
              if (sol[j] == head) nn = j;
            continue;
          }
          cwk2 = cswk;
          nn = 0;
          if (sol[j] == head) nn = j;
        }
      }
    }
    if (nn != 0) {
      work2[i] = cwk2;
      sol[i] = nn;
      sol[nn] = i;
    }
  }
// initial labeling
nn = 0;
for (i=1; i<=n; i++)
  if (sol[i] == head) {
    nn++;
    aux6[i] = 0;
    work4[i] = 0.;
    xwk2 = work2[i];
    for (j=1; j<=n; j++) {
      min = i;
      max = j;
      if (i != j) {
        if (i > j) {
          max = i;
          min = j;
        }
        sub = aux1[max] + min;
        xcst = cost[sub];
        cswk = cost[sub] - xwk2 - work2[j];
        if (cswk < work1[j]) {
          work1[j] = cswk;
          aux4[j] = i;
        }
      }
    }
  }
```

```
if (nn <= 1) fin = true;
// examine the labeling and prepare for the next step
iterate:
while (true) {
  if (fin) {
    // generate the original graph by expanding all shrunken blossoms
    skip = false;
    value = 0.;
    for (i=1; i<=n; i++)
      if (aux2[i] == i) {
        if (aux6[i] >= 0) {
          kk5 = sol[i];
          kk2 = aux2[kk5];
          kk4 = sol[kk2];
          aux6[i] = -1;
          aux6[kk2] = -1;
          min = kk4;
          max = kk5;
          if (kk4 != kk5) {
            if (kk4 > kk5) {
              max = kk4;
              min = kk5;
            }
            sub = aux1[max] + min;
            xcst = cost[sub];
            value += xcst;
          }
        }
      }
    for (i=1; i<=n; i++) {
      while (true) {
        idxb = aux2[i];
        if (idxb == i) break;
        mm2 = aux3[idxb];
        idxd = aux4[mm2];
        kk3 = mm2;
        xwork = work4[mm2];
        do {
          mm1 = mm2;
          idxe = aux5[mm1];
          xwk2 = work2[mm1];
          while (true) {
            aux2[mm2] = mm1;
            work3[mm2] -= xwk2;
            if (mm2 == idxe) break;
            mm2 = aux3[mm2];
          }
          mm2 = aux3[idxe];
          aux3[idxe] = mm1;
        } while (mm2 != idxd);
```

```
work2[idxb] = xwork;
aux3[idxb] = idxd;
mm2 = idxd;
while (true) {
  work3[mm2] -= xwork;
  if (mm2 == idxb) break;
  mm2 = aux3[mm2];
}
mm5 = sol[idxb];
mm1 = aux2[mm5];
mm1 = sol[mm1];
kk1 = aux2[mm1];
if (idxb != kk1) {
  sol[kk1] = mm5;
  kk3 = aux7[kk1];
  kk3 = aux2[kk3];
  do {
    mm3 = aux6[kk1];
    kk2 = aux2[mm3];
    mm1 = aux7[kk2];
    mm2 = aux8[kk2];
    kk1 = aux2[mm1];
    sol[kk1] = mm2;
    sol[kk2] = mm1;
    min = mm1;
    max = mm2;
    if (mm1 == mm2) {
      skip = true;
      break;
    }
    if (mm1 > mm2) {
      max = mm1;
      min = mm2;
    }
    sub = aux1[max] + min;
    xcst = cost[sub];
    value += xcst;
  } while (kk1 != idxb);
  if (kk3 == idxb) skip = true;
}
if (skip)
  skip = false;
else {
  while (true) {
    kk5 = aux6[kk3];
    kk2 = aux2[kk5];
    kk6 = aux6[kk2];
    min = kk5;
    max = kk6;
    if (kk5 == kk6) break;
```

```
          if (kk5 > kk6) {
            max = kk5;
            min = kk6;
          }
          sub = aux1[max] + min;
          xcst = cost[sub];
          value += xcst;
          kk6 = aux7[kk2];
          kk3 = aux2[kk6];
          if (kk3 == idxb) break;
        }
      }
    }
  }
  weight[0][0] = value;
  return;
}
cstlow = big;
for (i=1; i<=n; i++)
  if (aux2[i] == i) {
    cst = work1[i];
    if (aux6[i] < head) {
      cst = 0.5 * (cst + work4[i]);
      if (cst <= cstlow) {
        index = i;
        cstlow = cst;
      }
    }
    else {
      if (aux7[i] < head) {
        if (aux3[i] !- i) {
          cst += work2[i];
          if (cst < cstlow) {
            index = i;
            cstlow = cst;
          }
        }
      }
      else {
        if (cst < cstlow) {
          index = i;
          cstlow - cst;
        }
      }
    }
  }
if (aux7[index] >= head) {
  skip = false;
  if (aux6[index] < head) {
    idxd = aux4[index];
```

```
idxe = aux5[index];
kk4 = index;
kk1 = kk4;
kk5 = aux2[idxd];
kk2 = kk5;
while (true) {
  aux7[kk1] = kk2;
  mm5 = aux6[kk1];
  if (mm5 == 0) break;
  kk2 = aux2[mm5];
  kk1 = aux7[kk2];
  kk1 = aux2[kk1];
}
idxb = kk1;
kk1 = kk5;
kk2 = kk4;
while (true) {
  if (aux7[kk1] < head) break;
  aux7[kk1] = kk2;
  mm5 = aux6[kk1];
  if (mm5 == 0) {
    // augmentation of the matching
    // exchange the matching and non-matching edges
    //   along the augmenting path
    idxb = kk4;
    mm5 = idxd;
    while (true) {
      kk1 = idxb;
      while (true) {
        sol[kk1] = mm5;
        mm5 = aux6[kk1];
        aux7[kk1] = head;
        if (mm5 == 0) break;
        kk2 = aux2[mm5];
        mm1 = aux7[kk2];
        mm5 = aux8[kk2];
        kk1 = aux2[mm1];
        sol[kk2] = mm1;
      }
      if (idxb != kk4) break;
      idxb = kk5;
      mm5 = idxe;
    }
    // remove all labels on on-exposed base nodes
    for (i=1; i<=n; i++)
      if (aux2[i] == i) {
        if (aux6[i] < head) {
          cst = cstlow - work4[i];
          work2[i] += cst;
          aux6[i] = head;
```

```
          if (sol[i] != head)
            work4[i] = big;
          else {
            aux6[i] = 0;
            work4[i] = 0.;
          }
        }
        else {
          if (aux7[i] < head) {
            cst = work1[i] - cstlow;
            work2[i] += cst;
            aux7[i] = head;
            aux8[i] = head;
          }
          work4[i] = big;
        }
        work1[i] = big;
      }
    nn -= 2;
    if (nn <= 1) {
      fin = true;
      continue iterate;
    }
    // determine the new work1 values
    for (i=1; i<=n; i++) {
      kk1 = aux2[i];
      if (aux6[kk1] == 0) {
        xwk2 = work2[kk1];
        xwk3 = work3[i];
        for (j=1; j<=n; j++) {
          kk2 = aux2[j];
          if (kk1 != kk2) {
            min = i;
            max = j;
            if (i != j) {
              if (i > j) {
                max = i;
                min = j;
              }
              sub = aux1[max] + min;
              xcst = cost[sub];
              cswk = cost[sub] - xwk2 - xwk3;
              cswk -= (work2[kk2] + work3[j]);
              if (cswk < work1[kk2]) {
                aux4[kk2] = i;
                aux5[kk2] = j;
                work1[kk2] = cswk;
              }
            }
          }
```

```
            }
          }
        }
        continue iterate;
      }
      kk2 = aux2[mm5];
      kk1 = aux7[kk2];
      kk1 = aux2[kk1];
    }
    while (true) {
      if (kk1 == idxb) {
        skip = true;
        break;
      }
      mm5 = aux7[idxb];
      aux7[idxb] = head;
      idxa = sol[mm5];
      idxb = aux2[idxa];
    }
  }
  if (!skip) {
    // growing an alternating tree, add two edges
    aux7[index] = aux4[index];
    aux8[index] = aux5[index];
    idxa = sol[index];
    idxc = aux2[idxa];
    work4[idxc] = cstlow;
    aux6[idxc] = sol[idxc];
    msmSubprogramb(idxc,n,big,cost,aux1,aux2,aux3,aux4,
                   aux5,aux7,aux9,work1,work2,work3,work4);
    continue;
  }
  skip = false;
  // shrink a blossom
  xwork = work2[idxb] + cstlow - work4[idxb];
  work2[idxb] = 0.;
  mm1 = idxb;
  do {
    work3[mm1] += xwork;
    mm1 = aux3[mm1];
  } while (mm1 != idxb);
  mm5 = aux3[idxb];
  if (idxb == kk5) {
    kk5 = kk4;
    kk2 = aux7[idxb];
  }
  while (true) {
    aux3[mm1] = kk2;
    idxa = sol[kk2];
    aux6[kk2] = idxa;
```

```
        xwk2 = work2[kk2] + work1[kk2] - cstlow;
        mm1 = kk2;
        do {
          mm2 = mm1;
          work3[mm2] += xwk2;
          aux2[mm2] = idxb;
          mm1 = aux3[mm2];
        } while (mm1 != kk2);
        aux5[kk2] = mm2;
        work2[kk2] = xwk2;
        kk1 = aux2[idxa];
        aux3[mm2] = kk1;
        xwk2 = work2[kk1] + cstlow - work4[kk1];
        mm2 = kk1;
        do {
          mm1 = mm2;
          work3[mm1] += xwk2;
          aux2[mm1] = idxb;
          mm2 = aux3[mm1];
        } while (mm2 != kk1);
        aux5[kk1] = mm1;
        work2[kk1] = xwk2;
        if (kk5 != kk1) {
          kk2 = aux7[kk1];
          aux7[kk1] = aux8[kk2];
          aux8[kk1] = aux7[kk2];
          continue;
        }
        if (kk5 != index) {
          aux7[kk5] = idxe;
          aux8[kk5] = idxd;
          if (idxb != index) {
            kk5 = kk4;
            kk2 = aux7[idxb];
            continue;
          }
        }
        else {
          aux7[index] = idxd;
          aux8[index] = idxe;
        }
        break;
      }
      aux3[mm1] = mm5;
      kk4 = aux3[idxb];
      aux4[kk4] = mm5;
      work4[kk4] = xwork;
      aux7[idxb] = head;
      work4[idxb] = cstlow;
      msmSubprogramb(idxb,n,big,cost,aux1,aux2,aux3,aux4,
```

```
                    aux5,aux7,aux9,work1,work2,work3,work4);
    continue iterate;
  }
  // expand a t-labeled blossom
  kk4 = aux3[index];
  kk3 = kk4;
  idxd = aux4[kk4];
  mm2 = kk4;
  do {
    mm1 = mm2;
    idxe = aux5[mm1];
    xwk2 = work2[mm1];
    while (true) {
      aux2[mm2] = mm1;
      work3[mm2] -= xwk2;
      if (mm2 == idxe) break;
      mm2 = aux3[mm2];
    }
    mm2 = aux3[idxe];
    aux3[idxe] = mm1;
  } while (mm2 != idxd);
  xwk2 = work4[kk4];
  work2[index] = xwk2;
  aux3[index] = idxd;
  mm2 = idxd;
  while (true) {
    work3[mm2] -= xwk2;
    if (mm2 == index) break;
    mm2 = aux3[mm2];
  }
  mm1 = sol[index];
  kk1 = aux2[mm1];
  mm2 = aux6[kk1];
  idxb = aux2[mm2];
  if (idxb != index) {
    kk2 = idxb;
    while (true) {
      mm5 = aux7[kk2];
      kk1 = aux2[mm5];
      if (kk1 == index) break;
      kk2 = aux6[kk1];
      kk2 = aux2[kk2];
    }
    aux7[idxb] = aux7[index];
    aux7[index] = aux8[kk2];
    aux8[idxb] = aux8[index];
    aux8[index] = mm5;
    mm3 = aux6[idxb];
    kk3 = aux2[mm3];
    mm4 = aux6[kk3];
```

```
      aux6[idxb] = head;
      sol[idxb] = mm1;
      kk1 = kk3;
      while (true) {
        mm1 = aux7[kk1];
        mm2 = aux8[kk1];
        aux7[kk1] = mm4;
        aux8[kk1] = mm3;
        aux6[kk1] = mm1;
        sol[kk1] = mm1;
        kk2 = aux2[mm1];
        sol[kk2] = mm2;
        mm3 = aux6[kk2];
        aux6[kk2] = mm2;
        if (kk2 == index) break;
        kk1 = aux2[mm3];
        mm4 = aux6[kk1];
        aux7[kk2] = mm3;
        aux8[kk2] = mm4;
      }
    }
    mm2 = aux8[idxb];
    kk1 = aux2[mm2];
    work1[kk1] = cstlow;
    kk4 = 0;
    skip = false;
    if (kk1 != idxb) {
      mm1 = aux7[kk1];
      kk3 = aux2[mm1];
      aux7[kk1] = aux7[idxb];
      aux8[kk1] = mm2;
      do {
        mm5 = aux6[kk1];
        aux6[kk1] = head;
        kk2 = aux2[mm5];
        mm5 = aux7[kk2];
        aux7[kk2] = head;
        kk5 = aux8[kk2];
        aux8[kk2] = kk4;
        kk4 = kk2;
        work4[kk2] = cstlow;
        kk1 = aux2[mm5];
        work1[kk1] = cstlow;
      } while (kk1 != idxb);
      aux7[idxb] = kk5;
      aux8[idxb] = mm5;
      aux6[idxb] = head;
      if (kk3 == idxb) skip = true;
    }
    if (skip)
```

```
        skip = false;
      else {
        kk1 = 0;
        kk2 = kk3;
        do {
          mm5 = aux6[kk2];
          aux6[kk2] = head;
          aux7[kk2] = head;
          aux8[kk2] = kk1;
          kk1 = aux2[mm5];
          mm5 = aux7[kk1];
          aux6[kk1] = head;
          aux7[kk1] = head;
          aux8[kk1] = kk2;
          kk2 = aux2[mm5];
        } while (kk2 != idxb);
        msmSubprograma(kk1,n,big,cost,aux1,aux2,aux3,aux4,aux5,
                       aux6,aux8,work1,work2,work3,work4);
      }
      while (true) {
        if (kk4 == 0) continue iterate;
        idxb = kk4;
        msmSubprogramb(idxb,n,big,cost,aux1,aux2,aux3,aux4,
                       aux5,aux7,aux9,work1,work2,work3,work4);
        kk4 = aux8[idxb];
        aux8[idxb] = head;
      }
    }
}


static private void msmSubprograma(int kk, int n, double big,
           double cost[], int aux1[], int aux2[], int aux3[],
           int aux4[], int aux5[], int aux6[], int aux8[],
           double work1[], double work2[], double work3[],
           double work4[])
{
  /* this method is used internally by minSumMatching */

  int i,head,j,jj1,jj2,jj3,jj4,min,max,sub;
  double cswk,cstwk,xcst,xwk2,xwk3;

  head = n + 2;
  do {
    jj1 = kk;
    kk = aux8[jj1];
    aux8[jj1] = head;
    cstwk = big;
    jj3 = 0;
    jj4 = 0;
```

```
    j = jj1;
    xwk2 = work2[jj1];
    do {
      xwk3 = work3[j];
      for (i=1; i<=n; i++) {
        jj2 = aux2[i];
        if (aux6[jj2] < head) {
          min = j;
          max = i;
          if (j != i) {
            if (j > i) {
              max = j;
              min = i;
            }
            sub = aux1[max] + min;
            xcst = cost[sub];
            cswk = cost[sub] - xwk2 - xwk3;
            cswk -= (work2[jj2] + work3[i]);
            cswk += work4[jj2];
            if (cswk < cstwk) {
              jj3 = i;
              jj4 = j;
              cstwk = cswk;
            }
          }
        }
      }
      j = aux3[j];
    } while (j != jj1);
    aux4[jj1] = jj3;
    aux5[jj1] = jj4;
    work1[jj1] = cstwk;
  } while (kk != 0);
}


static private void msmSubprogramb(int kk, int n, double big,
           double cost[], int aux1[], int aux2[], int aux3[],
           int aux4[], int aux5[], int aux7[], int aux9[],
           double work1[], double work2[], double work3[],
           double work4[])
{
  /* this method is used internally by minSumMatching */

  int i,ii,head,jj1,jj2,jj3,min,max,sub;
  double cswk,xcst,xwk1,xwk2;

  head = n + 2;
  xwk1 = work4[kk] - work2[kk];
  work1[kk] = big;
```

```
xwk2 = xwk1 - work3[kk];
aux7[kk] = 0;
ii = 0;
for (i=1; i<=n; i++) {
  jj3 = aux2[i];
  if (aux7[jj3] >= head) {
    ii++;
    aux9[ii] = i;
    min = kk;
    max = i;
    if (kk != i) {
      if (kk > i) {
        max = kk;
        min = i;
      }
      sub = aux1[max] + min;
      cswk = cost[sub] + xwk2;
      cswk -= (work2[jj3] + work3[i]);
      if (cswk < work1[jj3]) {
        aux4[jj3] = kk;
        aux5[jj3] = i;
        work1[jj3] = cswk;
      }
    }
  }
}
aux7[kk] = head;
jj1 = kk;
jj1 = aux3[jj1];
if (jj1 == kk) return;
do {
  xwk2 = xwk1 - work3[jj1];
  for (i=1; i<=ii; i++) {
    jj2 = aux9[i];
    jj3 = aux2[jj2];
    min = jj1;
    max = jj2;
    if (jj1 != jj2) {
      if (jj1 > jj2) {
        max = jj1;
        min = jj2;
      }
      sub = aux1[max] + min;
      xcst = cost[sub];
      cswk = cost[sub] + xwk2;
      cswk -= (work2[jj3] + work3[jj2]);
      if (cswk < work1[jj3]) {
        aux4[jj3] = jj1;
        aux5[jj3] = jj2;
        work1[jj3] = cswk;
```

```
        }
      }
    }
    jj1 = aux3[jj1];
  } while (jj1 != kk);
}
```

Example:

Find the minimum cost perfect matching of an undirected graph of 8 vertices with the following given weight matrix:

$$\begin{pmatrix} 0 & 24 & 16 & 18 & 17 & 16 & 23 & 25 \\ & 0 & 15 & 22 & 24 & 23 & 18 & 16 \\ & & 0 & 24 & 17 & 16 & 25 & 24 \\ & & & 0 & 22 & 23 & 24 & 15 \\ & & & & 0 & 24 & 15 & 16 \\ & & & & & 0 & 18 & 17 \\ & & & & & & 0 & 25 \\ & & & & & & & 0 \end{pmatrix}$$

```
package GraphAlgorithms;

public class Test_minSumMatching extends Object {

  public static void main(String args[]) {
    int n = 8;
    double weight[][] = {{0., 0., 0., 0., 0., 0., 0., 0., 0.},
                         {0., 0.,24.,16.,18.,17.,16.,23.,25.},
                         {0., 0., 0.,15.,22.,24.,23.,18.,16.},
                         {0., 0., 0., 0.,24.,17.,16.,25.,24.},
                         {0., 0., 0., 0., 0.,22.,23.,24.,15.},
                         {0., 0., 0., 0., 0., 0.,24.,15.,16.},
                         {0., 0., 0., 0., 0., 0., 0.,18.,17.},
                         {0., 0., 0., 0., 0., 0., 0., 0.,25.},
                         {0., 0., 0., 0., 0., 0., 0., 0., 0.}};
    int sol[] = new int[n + 1];

    GraphAlgo.minSumMatching(n, weight, sol);
    System.out.println("Optimal matching:");
    for (int i=1; i<=n; i++)
      System.out.println("  " + i + " -- " + sol[i]);
    System.out.println("\nTotal optimal matching cost = " + weight[0][0]);
  }
}
```

Output:

```
Optimal matching:
  1 -- 6
  2 -- 3
  3 -- 2
  4 -- 8
  5 -- 7
  6 -- 1
  7 -- 5
  8 -- 4

Total optimal matching cost = 61.0
```

8. Network Flow

8.1 Maximum Network Flow

A *network* is a direct graph G in which each edge (p,q) is associated with a nonnegative number c(p,q) called the *capacity* of the edge. Let the number f(p,q) be the *flow* from node p to node q. Let *source* and *sink* be two specified nodes. A flow in the network is *feasible* if f(p,q) does not exceed c(p,q) for each edge (p,q) in G, and for every node p other than the source and sink, the sum of all flows incoming to node p is equal to the sum of all flows outgoing from node p. The *maximum network flow problem* is to find a feasible flow in the network from the source to the sink such that the amount of flow into the sink is maximum.

A *cut* is a subset S of the nodes of G with the capacity equal to:

$$\sum_{\substack{i \in S \\ j \notin S}} c(p,q)$$

The well-known *max-flow min-cut theorem* [FF56] states that the maximum flow is equal to the minimum cut in a network. The following preflow algorithm [E79, K74] finds a maximum flow and a minimum cut set in a given network of *n* nodes with a running time $O(n^3)$.

For a given source node s and sink node t, the algorithm starts with an initial preflow where all edges from s have zero residual capacity, producing excesses at the adjacent nodes of s. The excesses are pushed towards t, while satisfying the constraint that the in-flow must equal the out-flow at each node. The excesses which cannot reach the sink due to capacity constraints are returned back to s. The algorithm terminates when there are no more excesses.

Procedure parameters:

void maximumNetworkFlow(*n, m, nodei, nodej, capacity, source, sink, minimumcut, arcflow, nodeflow*)

n: int;
 entry: the number of nodes of the network, labeled from 1 to *n*.
m: int;
 entry: the number of edges in the network.
nodei, nodej: int[2*m+1];
 entry: *nodei[p]* and *nodej[p]* are the end nodes of the p-th edge in the graph, p=1,2,...,*m*. Other elements of the two arrays are not required in the input.
 exit: the edges (*nodei[p], nodej[p]*), p=1,2,...,m, will be sorted lexicographically.

capacity: int[2*m+1];
entry: *capacity[p]* is the capacity of edge p in the graph, p=1,2,...,m. Other elements of this array are not required in the input.
exit: the original content of the array *capacity* will be destroyed.

source: int;
entry: the source node of the graph, $1 \le source \le n$.

sink: int;
entry: the sink node of the graph, $1 \le sink \le n$.

minimumcut: int[n+1];
exit: *minimumcut[i]* = 1 if node i is in the minimum cut set, otherwise *minimumcut[i]* = 0, for i = 1, 2, ..., n.

arcflow: int[m+1];
exit: *arcflow[0]* is the number of edges that have nonzero flow in the result, *arcflow[0]* $\le m$. *arcflow[i]* is the amount of flow through the i-th edge, i=1,2,...,*arcflow[0]*.

nodeflow: int[n+1];
exit: *nodeflow[i]* is the amount of flow through node i, i=1,2,...,*arcflow[0]*. *nodeflow[0]* is the total amount of flow.

```
public static void maximumNetworkFlow(int n, int m, int nodei[],
              int nodej[], int capacity[], int source, int sink,
              int minimumcut[], int arcflow[], int nodeflow[])
{
  int i,j,curflow,flag,medge,nodew,out;
  int in=0,iout=0,parm=0,m1=0,icont=0,jcont=0;
  int last=0,nodep=0,nodeq=0,nodeu=0,nodev=0,nodex=0,nodey=0;
  int firstarc[] = new int[n+1];
  int imap[] = new int[n+1];
  int jmap[] = new int[n+1];
  boolean finish,controla,controlb,controlc,controlg;
  boolean controld=false,controle=false,controlf=false;

  // create the artificial edges
  j = m;
  for (i=1; i<=m; i++) {
    j++;
    nodei[m+i] = nodej[i];
    nodej[m+i] = nodei[i];
    capacity[m+i] = 0;
  }
  m = m + m;
  // initialize
  for (i=1; i<=n; i++)
    firstarc[i] = 0;
  curflow = 0;
  for (i=1; i<=m; i++) {
    arcflow[i] = 0;
    j = nodei[i];
    if (j == source) curflow += capacity[i];
    firstarc[j]++;
```

```
}
nodeflow[source] = curflow;
nodew = 1;
for (i=1; i<=n; i++) {
  j = firstarc[i];
  firstarc[i] = nodew;
  imap[i] = nodew;
  nodew += j;
}
finish = false;
controla = true;
// sort the edges in lexicographical order
entry1:
while (true) {
  flag = 0;
  controlb = false;
  entry2:
  while (true) {
    if (!controlb) {
      if ((flag < 0) && controla) {
        if (flag != -1) {
          if (nodew < 0) nodep++;
          nodeq = jcont;
          jcont = nodep;
          flag = -1;
        }
        else {
          if (nodew <= 0) {
            if (icont > 1) {
              icont--;
              jcont = icont;
              controla = false;
              continue entry2;
            }
            if (m1 == 1)
              flag = 0;
            else {
              nodep = m1;
              m1--;
              nodeq = 1;
              flag = 1;
            }
          }
          else
            flag = 2;
        }
      }
      else {
        if (controla)
          if (flag > 0) {
```

```
          if (flag <= 1) jcont = icont;
          controla = false;
        }
      if (controla) {
        m1 = m;
        icont = 1 + m / 2;
        icont--;
        jcont = icont;
      }
      controla = true;
      nodep = jcont + jcont;
      if (nodep < m1) {
        nodeq = nodep + 1;
        flag = -2;
      }
      else {
        if (nodep == m1) {
          nodeq = jcont;
          jcont = nodep;
          flag = -1;
        }
        else {
          if (icont > 1) {
            icont--;
            jcont = icont;
            controla = false;
            continue entry2;
          }
          if (m1 == 1)
            flag = 0;
          else {
            nodep = m1;
            m1--;
            nodeq = 1;
            flag = 1;
          }
        }
      }
    }
  }
  controlg = false;
  controlc = false;
  if ((flag < 0) && !controlb) {
     nodew = nodei[nodep] - nodei[nodeq];
     if (nodew == 0) nodew = nodej[nodep] - nodej[nodeq];
     continue entry2;
  }
  else {
    if ((flag > 0) || controlb) {
      // interchange two edges
```

```
      controlb = false;
      nodew = nodei[nodep];
      nodei[nodep] = nodei[nodeq];
      nodei[nodeq] = nodew;
      curflow = capacity[nodep];
      capacity[nodep] = capacity[nodeq];
      capacity[nodeq] = curflow;
      nodew = nodej[nodep];
      nodej[nodep] = nodej[nodeq];
      nodej[nodeq] = nodew;
      curflow = arcflow[nodep];
      arcflow[nodep] = arcflow[nodeq];
      arcflow[nodeq] = curflow;
      if (flag > 0) continue entry2;
      if (flag == 0) {
        controlc = true;
      }
      else {
        jmap[nodev] = nodeq;
        controlg = true;
      }
    }
    else
      if (finish) {
        // return the maximum flow on each edge
        j = 0;
        for (i=1; i<=m; i++)
          if (arcflow[i] > 0) {
            j++;
            nodei[j] = nodei[i];
            nodej[j] = nodej[i];
            arcflow[j] = arcflow[i];
          }
        arcflow[0] = j;
        return;
      }
  }
  if (!controlg && !controlc) {
    // set the cross references between edges
    for (i=1; i<=m; i++) {
      nodev = nodej[i];
      nodei[i] = imap[nodev];
      imap[nodev]++;
    }
  }
  entry3:
  while (true) {
    if (!controlg) {
      if (!controlc) {
        flag = 0;
```

```
      for (i=1; i<=n; i++) {
        if (i != source) nodeflow[i] = 0;
        jmap[i] = m + 1;
        if (i < n) jmap[i] = firstarc[i + 1];
        minimumcut[i] = 0;
      }
      in = 0;
      iout = 1;
      imap[1] = source;
      minimumcut[source] = -1;
      while (true) {
        in++;
        if (in > iout) break;
        nodeu = imap[in];
        medge = jmap[nodeu] - 1;
        last = firstarc[nodeu] - 1;
        while (true) {
          last++;
          if (last > medge) break;
          nodev = nodej[last];
          curflow = capacity[last] - arcflow[last];
          if ((minimumcut[nodev] != 0) || (curflow == 0)) continue;
          if (nodev != sink) {
            iout++;
            imap[iout] = nodev;
          }
          minimumcut[nodev] = -1;
        }
      }
      if (minimumcut[sink] == 0) {
        // exit
        for (i=1; i<=n; i++)
          minimumcut[i] = -minimumcut[i];
        for (i=1; i<=m; i++) {
          nodeu = nodej[nodei[i]];
          if (arcflow[i] < 0) nodeflow[nodeu] -= arcflow[i];
          nodei[i] = nodeu;
        }
        nodeflow[source] = nodeflow[sink];
        finish = true;
        continue entry1;
      }
      minimumcut[sink] = 1;
    }
    while (true) {
      if (!controlc) {
        in--;
        if (in == 0) break;
        nodeu = imap[in];
        nodep = firstarc[nodeu] - 1;
```

```
          nodeq = jmap[nodeu] - 1;
        }
        controlc = false;
        while (nodep != nodeq) {
          nodev = nodej[nodeq];
          if ((minimumcut[nodev] <= 0) ||
              (capacity[nodeq] == arcflow[nodeq])) {
            nodeq--;
            continue;
          }
          else {
            nodej[nodeq] = -nodev;
            capacity[nodeq] -= arcflow[nodeq];
            arcflow[nodeq] = 0;
            nodep++;
            if (nodep < nodeq) {
              nodei[nodei[nodep]] = nodeq;
              nodei[nodei[nodeq]] = nodep;
              controlb = true;
              continue entry2;
            }
            break;
          }
        }
        if (nodep >= firstarc[nodeu]) minimumcut[nodeu] = nodep;
      }
      nodex = 0;
      for (i=1; i<=iout; i++)
        if (minimumcut[imap[i]] > 0) {
          nodex++;
          imap[nodex] = imap[i];
        }
      // find a feasible flow
      flag = -1;
      nodey = 1;
    }
    entry4:
    while (true) {
      if (!controlg) {
        if (!controlf) {
          if (!controld && !controle)
            nodeu = imap[nodey];
          if ((nodeflow[nodeu] <= 0) || controld || controle) {
            if (!controle) {
              controld = false;
              nodey++;
              if (nodey <= nodex) continue entry4;
              parm = 0;
            }
            controle = false;
```

```
nodey--;
if (nodey != 1) {
  nodeu = imap[nodey];
  if (nodeflow[nodeu] < 0) {
    controle = true;
    continue entry4;
  }
  if (nodeflow[nodeu] == 0) {
    // accumulating flows
    medge = m + 1;
    if (nodeu < n) medge = firstarc[nodeu + 1];
    last = jmap[nodeu];
    jmap[nodeu] = medge;
    while (true) {
      if (last == medge) {
        controle = true;
        continue entry4;
      }
      j = nodei[last];
      curflow = arcflow[j];
      arcflow[j] = 0;
      capacity[j] -= curflow;
      arcflow[last] -= curflow;
      last++;
    }
  }
  if (firstarc[nodeu] > minimumcut[nodeu]) {
    last = jmap[nodeu];
    do {
      j = nodei[last];
      curflow = arcflow[j];
      if (nodeflow[nodeu] < curflow) curflow = nodeflow[nodeu];
      arcflow[j] -= curflow;
      nodeflow[nodeu] -= curflow;
      nodev = nodej[last];
      nodeflow[nodev] += curflow;
      last++;
    } while (nodeflow[nodeu] > 0);
    nodeflow[nodeu] = -1;
    controle = true;
    continue entry4;
  }
  last = minimumcut[nodeu] + 1;
  controlf = true;
  continue entry4;
}
for (i=1; i<=m; i++) {
  nodev = -nodej[i];
  if (nodev >= 0) {
    nodej[i] = nodev;
```

```
            j = nodei[i];
            capacity[i] -= arcflow[j];
            curflow = arcflow[i] - arcflow[j];
            arcflow[i] = curflow;
            arcflow[j] = -curflow;
          }
        }
        continue entry3;
      }
      // an outgoing edge from a node is given maximum flow
      last = minimumcut[nodeu] + 1;
    }
  }
  while (true) {
    if (!controlg) {
      controlf = false;
      last--;
      if (last < firstarc[nodeu]) break;
      nodev = -nodej[last];
      if (nodeflow[nodev] < 0) continue;
      curflow = capacity[last] - arcflow[last];
      if (nodeflow[nodeu] < curflow) curflow = nodeflow[nodeu];
      arcflow[last] += curflow;
      nodeflow[nodeu] -= curflow;
      nodeflow[nodev] += curflow;
      parm = 1;
      nodep = nodei[last];
      nodeq = jmap[nodev] - 1;
      if (nodep < nodeq) {
        nodei[nodei[nodep]] = nodeq;
        nodei[nodei[nodeq]] = nodep;
        controlb = true;
        continue entry2;
      }
      if (nodep == nodeq) jmap[nodev] = nodeq;
    }
    controlg = false;
    if (nodeflow[nodeu] > 0) continue;
    if (capacity[last] == arcflow[last]) last--;
    break;
  }
  minimumcut[nodeu] = last;
  if (parm != 0) {
    controld = true;
    continue entry4;
  }
  // remove excess incoming flows from nodes
  last = jmap[nodeu];
  do {
    j = nodei[last];
```

```
            curflow = arcflow[j];
            if (nodeflow[nodeu] < curflow) curflow = nodeflow[nodeu];
            arcflow[j] -= curflow;
            nodeflow[nodeu] -= curflow;
            nodev = nodej[last];
            nodeflow[nodev] += curflow;
            last++;
          } while (nodeflow[nodeu] > 0);
          nodeflow[nodeu] = -1;
          controle = true;
          continue entry4;
        }
      }
    }
  }
}
```

Example:

Find the maximum flow of the following network from the source node 5 to the sink node 2. A capacity is associated with each edge.

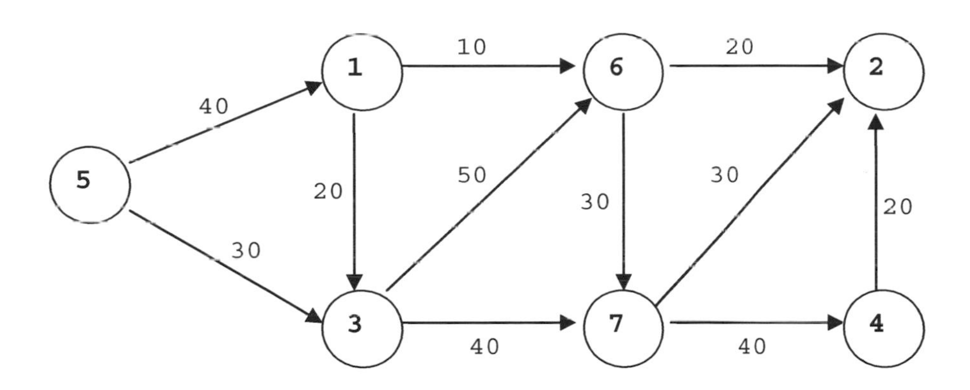


```
package GraphAlgorithms;

public class Test_maximumNetworkFlow extends Object {

  public static void main(String args[]) {

    int n=7, m=11;
    int nodei[]    = {0, 5, 5, 1, 1, 3, 3, 6, 7, 7, 6, 4,
                         0, 0, 0, 0, 0, 0, 0, 0, 0, 0, 0};
    int nodej[]    = {0, 3, 1, 3, 6, 7, 6, 7, 4, 2, 2, 2,
                         0, 0, 0, 0, 0, 0, 0, 0, 0, 0, 0};
    int capacity[] = {0,30,40,20,10,40,50,30,40,30,20,20,
                         0, 0, 0, 0, 0, 0, 0, 0, 0, 0, 0};
    int source=5, sink=2;
```

```
    int minimumcut[] = new int[n+1];
    int nodeflow[] = new int[n+1];
    int arcflow[] = new int[m+m+1];

    GraphAlgo.maximumNetworkFlow(n,m,nodei,nodej,capacity,source,sink,
                                minimumcut,arcflow,nodeflow);
    System.out.print("Nodes in the minimum cut set: ");
    for (int i=1; i<=n; i++)
      if (minimumcut[i] == 1)
        System.out.print("  " + i);
    System.out.println("\n\nAmount of flow through each node:" +
                       "\n  node  flow");
    for (int i=1; i<=n; i++)
      System.out.println("    " + i + "    " + nodeflow[i]);
    System.out.println("\nAmount of flow on each edge:\n  from to  flow");
    for (int i=1; i<=arcflow[0]; i++)
      System.out.printf("%5d%4d%5d\n",nodei[i],nodej[i],arcflow[i]);
  }
}
```

Output:

```
Nodes in the minimum cut set:   1  5

Amount of flow through each node:
  node  flow
    1    30
    2    60
    3    50
    4    10
    5    60
    6    50
    7    40

Amount of flow on each edge:
  from to  flow
    1   3   20
    1   6   10
    3   6   40
    3   7   10
    4   2   10
    5   1   30
    5   3   30
    6   2   20
    6   7   30
    7   2   30
    7   4   10
```

8.2 Minimum Cost Network Flow

Consider a directed graph, called a *network*, in which some nodes are sources with specified units of supply, and some nodes with specified units of demand. Other nodes which are neither sources nor sinks are called intermediate nodes. Each edge in the network is associated with a lower bound of required units of flow, an upper bound capacity, and a cost per unit flow. The *minimum cost network flow problem* is to send the units of supply from the sources to meet all the demand of the sinks along the edges while satisfying the edge lower and upper bounds at the minimum cost. The following primal simplex algorithm [KH80] solves the minimum cost network flow problem.

Procedure parameters:

int minCostNetworkFlow (nodes, *edges, numdemand, nodedemand, nodei, nodej, arccost, upbound, lowbound, arcsol, flowsol*)

minCostNetworkFlow: int;
  exit: the method returns an integer with the following values:
    0: optimal solution found
    1: Infeasible, net required flow is negative
    2: Need to increase the size of internal edge-length arrays
    3: Error in the input of the arc list, arc cost, and arc flow
    4: Infeasible, net required flow imposed by arc flow lower bounds is negative

nodes: int;
  entry: the number of nodes of the graph, labeled from 1 to *nodes*.

edges: int;
  entry: the number of edges in the graph.

numdemand: int;
  entry: the number nonzero flow requirement at the nodes.

nodedemand: int[*numdemand*+1][2];
  entry: *nodedemand[i][1]* is the flow requirement at node *nodedemand[i][0]*, for i = 1, 2, ..., *nodedemand*. The supply at a node will be positive flow requirement, and demand at a node will be negative flow requirement.

nodei, nodej: int[*edges*+1];
  entry: *nodei[p]* and *nodej[p]* are the end nodes of the p-th edge in the graph, p=1,2,...,*edges*.

arccost: int[*edges*+1];
  entry: *arccost[i]* is the cost on edge i, for i = 1, 2, ..., *edges*.
  exit: *arccost[0]* is the total cost of the optimal solution.

upbound: int[*edges*+1];
  entry: *upbound[i]* is the upper bound flow requirement of edge i, for i = 1, 2, ..., *edges*. A zero value of upper bound on an edge denotes that edge to be unbounded. A –1 value of upper bound on an edge denotes that edge to have an upper bound value of zero.

lowbound: int[*edges*+1];
entry: *lowbound[i]* is the lower bound flow requirement of edge i, for i = 1, 2, ..., *edges*.

arcsol: int[2][*edges*+1];
exit: *arcsol[0][0]* is the number of edges that has nonzero flow value in the optimal solution; edge i is from node *arcsol[0][i]* to node *arcsol[1][i]*, for i = 1, 2, ..., *arcsol[0][0]*.

flowsol: int[*edges*+1];
exit: *flowsol[i]* is the value of the flow on edge i in the optimal solution, for i = 1, 2, ..., *edges*.

```
public static int minCostNetworkFlow(int nodes, int edges, int numdemand,
              int nodedemand[][], int nodei[], int nodej[], int arccost[],
              int upbound[], int lowbound[], int arcsol[][], int flowsol[])
{
  int i, j, k, l, m, n, lastslackedge, solarc, temp, tmp, u, v, remain, rate;
  int arcnam, tedges, tedges1, nodes1, nodes2, nzdemand, value, valuez;
  int tail, ratez, tailz, trial, distdiff, olddist, treenodes, iterations;
  int right, point, part, jpart, kpart, spare, sparez, lead, otherend, sedge;
  int orig, load, curedge, p, q, r, vertex1, vertex2, track, spointer, focal;
  int newpr, newlead, artedge, maxint, artedge1, ipart, distlen;
  int after=0, other=0, left=0, newarc=0, newtail=0;
  int pred[] = new int[nodes + 2];
  int succ[] = new int[nodes + 2];
  int dist[] = new int[nodes + 2];
  int sptpt[] = new int[nodes + 2];
  int flow[] = new int[nodes + 2];
  int dual[] = new int[nodes + 2];
  int arcnum[] = new int[nodes + 1];
  int head[] = new int[edges * 2];
  int cost[] = new int[edges * 2];
  int room[] = new int[edges * 2];
  int least[] = new int[edges * 2];
  int rim[] = new int[3];
  int ptr[] = new int[3];
  boolean infeasible;
  boolean flowz=false, newprz=false, artarc=false, removelist=false;
  boolean partz=false, ipartout=false, newprnb=false;

  for (p=0; p<=nodes; p++)
    arcnum[p] = 0;
  maxint = 0;
  for (p=1; p<=edges; p++) {
    arcnum[nodej[p]]++;
    if (arccost[p] > 0) maxint += arccost[p];
    if (upbound[p] > 0) maxint += upbound[p];
  }
  artedge = 1;
```

```
artedge1 = artedge + 1;
tedges =  (edges * 2) - 2;
tedges1 = tedges + 1;
nodes1 = nodes + 1;
nodes2 = nodes + 2;
dual[nodes1] = 0;
for (p=1; p<=nodes1; p++) {
  pred[p] = 0;
  succ[p] = 0;
  dist[p] = 0;
  sptpt[p] = 0;
  flow[p] = 0;
}
head[artedge] = nodes1;
cost[artedge] - maxint;
room[artedge] = 0;
least[artedge] = 0;
remain = 0;
nzdemand = 0;
sedge = 0;

// initialize supply and demand lists
succ[nodes1] = nodes1;
pred[nodes1] = nodes1;
for (p=1; p<=numdemand; p++) {
  flow[nodedemand[p][0]] = nodedemand[p][1];
  remain += nodedemand[p][1];
  if (nodedemand[p][1] <= 0) continue;
  nzdemand++;
  dist[nodedemand[p][0]] = nodedemand[p][1];
  succ[nodedemand[p][0]] - succ[nodes1];
  succ[nodes1] = nodedemand[p][0];
}
if (remain < 0)  return 1;
for (p=1; p<=nodes; p++)
  dual[p] = arcnum[p];
i = 1;
j = artedge;
for (p=1; p<=nodes; p++) {
  i = -i;
  tmp = Math.max(1, dual[p]);
  if (j + tmp > tedges) return 2;
  dual[p] = (i >= 0 ? j+1 : -(j+1));
  for (q=1; q<=tmp; q++) {
    j++;
    head[j] = (i >= 0 ? p : -p);
    cost[j] = 0;
    room[j] = -maxint;
    least[j] = 0;
  }
```

```
    }

    // check for valid input data
    sedge = j + 1;
    if (sedge > tedges)  return 2;
    head[sedge] = (-i >= 0 ? nodes1 : -nodes1);
    valuez = 0;
    for (p=1; p<=edges; p++) {
      if ((nodei[p] > nodes) || (nodej[p] > nodes) || (upbound[p] >= maxint))
        return 3;
      if (upbound[p] == 0) upbound[p] = maxint;
      if (upbound[p] < 0) upbound[p] = 0;
      if ((lowbound[p] >= maxint) || (lowbound[p] < 0) ||
                          (lowbound[p] > upbound[p]))
        return 3;
      u = dual[nodej[p]];
      v = Math.abs(u);
      temp = (u >= 0 ? nodes1 : -nodes1);
      if ((temp ^ head[v]) <= 0) {
        sedge++;
        tmp = sedge - v;
        r = sedge;
        for (q=1; q<=tmp; q++) {
          temp = r - 1;
          head[r]  = head[temp];
          cost[r]  = cost[temp];
          room[r] = room[temp];
          least[r] = least[temp];
          r = temp;
        }
        for (q=nodej[p]; q<=nodes; q++)
          dual[q] += (dual[q] >= 0 ? 1 : -1);
      }

      // insert new edge
      head[v] = (u >= 0 ? nodei[p] : -nodei[p]);
      cost[v] = arccost[p];
      valuez += arccost[p] * lowbound[p];
      room[v] = upbound[p] - lowbound[p];
      least[v] = lowbound[p];
      flow[nodei[p]] -= lowbound[p];
      flow[nodej[p]] += lowbound[p];
      dual[nodej[p]] = (u >= 0 ? v+1 : -(v+1));
      sptpt[nodei[p]] = -1;
    }
    i = nodes1;
    k = artedge;
    l = 0;
    sedge--;
    for (p=artedge1; p<=sedge; p++) {
```

```
      j = head[p];
      if ((i ^ j) <= 0) {
        i = -i;
        l++;
        dual[l] = k + 1;
      }
      else
       if (Math.abs(j) == 1) continue;
      k++;
      if (k != p) {
        head[k]  = head[p];
        cost[k]  = cost[p];
        room[k]  = room[p];
        least[k] = least[p];
      }
    }
    sedge = k;
    if (sedge + Math.max(1,nzdemand) + 1 > tedges) return 2;
    // add regular slacks
    i = -head[sedge];
    focal = succ[nodes1];
    succ[nodes1] = nodes1;
    if (focal == nodes1) {
      sedge++;
      head[sedge]  = (i >= 0 ? nodes1 : -nodes1);
      cost[sedge]  = 0;
      room[sedge]  = -maxint;
      least[sedge] = 0;
    }
    else
      do {
        sedge++;
        head[sedge] = (i >= 0 ? focal : -focal);
        cost[sedge] = 0;
        room[sedge] = dist[focal];
        dist[focal] = 0;
        least[sedge] = 0;
        after = succ[focal];
        succ[focal] = 0;
        focal = after;
      } while (focal != nodes1);
    lastslackedge = sedge;
    sedge++;
    head[sedge] = (-i >= 0 ? nodes2 : -nodes2);
    cost[sedge] = maxint;
    room[sedge] = 0;
    least[sedge] = 0;
    // locate sources and sinks
    remain = 0;
    treenodes = 0;
```

```
focal = nodes1;
for (p=1; p<=nodes; p++) {
  j = flow[p];
  remain += j;
  if (j == 0) continue;
  if (j < 0) {
    flow[p] = -j;
    right = nodes1;
    do {
      after = pred[right];
      if (flow[after]+j <= 0) break;
      right = after;
    } while (true);
    pred[right] = p;
    pred[p] = after;
    dist[p] = -1;
  }
  else {
    treenodes++;
    sptpt[p] = -sedge;
    flow[p] = j;
    succ[focal] = p;
    pred[p] = nodes1;
    succ[p] = nodes1;
    dist[p] = 1;
    dual[p] = maxint;
    focal = p;
  }
}
if (remain < 0) return 4;
do {
  // select highest rank demand
  tail = pred[nodes1];
  if (tail == nodes1) break;
  do {
    // set link to artificial
    newarc = artedge;
    newpr = maxint;
    newprz = false;
    flowz = false;
    if (flow[tail] == 0) {
      flowz = true;
      break;
    }
    // look for sources
    trial = dual[tail];
    lead = head[trial];
    other = (lead >= 0 ? nodes1 : -nodes1);
    do {
      if (room[trial] > 0) {
```

```
        orig = Math.abs(lead);
        if (dist[orig] == 1) {
          if (sptpt[orig] != artedge) {
            rate = cost[trial];
            if (rate < newpr) {
              if (room[trial] <= flow[tail]) {
                if (flow[orig] >= room[trial]) {
                  newarc = -trial;
                  newpr = rate;
                  if (newpr == 0) {
                    newprz = true;
                    break;
                  }
                }
              }
              else {
                if (flow[orig] >= flow[tail]) {
                  newarc = trial;
                  newpr = rate;
                  if (newpr == 0) {
                    newprz = true;
                    break;
                  }
                }
              }
            }
          }
        }
      }
      trial++;
      lead = head[trial];
    } while ((lead ^ other) > 0);
    if (!newprz) {
      artarc = false;
      if (newarc == artedge) {
        artarc = true;
        break;
      }
    } else
      newprz = false;
    if (newarc > 0) break;
    newarc = -newarc;
    orig = Math.abs(head[newarc]);
    load = room[newarc];
    // mark unavailable
    room[newarc] = -load;
    // adjust flows
    flow[orig] -= load;
    flow[tail] -= load;
  } while (true);
```

```
if (!flowz) {
  removelist = false;
  if (!artarc) {
    room[newarc] = -room[newarc];
    orig = Math.abs(head[newarc]);
    flow[orig] -= flow[tail];
    k = maxint;
    removelist = true;
  }
  else {
    // search for transshipment nodes
    artarc = false;
    trial = dual[tail];
    lead = head[trial];
    newprz = false;
    do {
      if (room[trial] > 0) {
        orig = Math.abs(lead);
        // is it linked
        if (dist[orig] == 0) {
          rate = cost[trial];
          if (rate < newpr) {
            newarc = trial;
            newpr = rate;
            if (newpr == 0) {
              newprz = true;
              break;
            }
          }
        }
      }
      trial++;
      lead = head[trial];
    } while ((lead ^ other) > 0);
    artarc = false;
    if (!newprz) {
      if(newarc == artedge)
        artarc = true;
    }
    else
      newprz = false;
    if (!artarc) {
      orig = Math.abs(head[newarc]);
      if (room[newarc] <= flow[tail]) {
        // get capacity
        load = room[newarc];
        // mark unavailable
        room[newarc] = -load;
        // adjust flows
        flow[orig] = load;
```

```
            flow[tail] -= load;
            pred[orig] = tail;
            pred[nodes1] = orig;
            dist[orig] = -1;
            continue;
          }
          // mark unavailable
          room[newarc] = -room[newarc];
          flow[orig] = flow[tail];
          pred[orig] = pred[tail];
          pred[tail] = orig;
          pred[nodes1] = orig;
          succ[orig] = tail;
          sptpt[tail] = newarc;
          dist[orig] = dist[tail] - 1;
          dual[tail] = newpr;
          treenodes++;
          continue;
        }
        else
          artarc = false;
      }
    }
    flowz = false;
    if (!removelist)
      k = 0;
    else
      removelist = false;
    pred[nodes1] = pred[tail];
    orig = Math.abs(head[newarc]);
    sptpt[tail] = newarc;
    dual[tail] = newpr;
    pred[tail] = orig;
    i = succ[orig];
    succ[orig] = tail;
    j = dist[orig] - dist[tail] + 1;
    focal = orig;
    do {
      // adjust dual variables
      focal = succ[focal];
      l = dist[focal];
      dist[focal] = l + j;
      k -= dual[focal];
      dual[focal] = k;
    } while (l != -1);
    succ[focal] = i;
    treenodes++;
  } while (true);

  // set up the expand tree
```

```
tail = 1;
trial = artedge1;
lead = head[trial];
do {
  if (treenodes == nodes) break;
  tailz = tail;
  newpr = maxint;
  do {
    // search for least cost connectable edge
    otherend = dist[tail];
    other = (lead >= 0 ? nodes1 : -nodes1);
    do {
      if (room[trial] > 0) {
        m = cost[trial];
        if (newpr >= m) {
          orig = Math.abs(lead);
          if (dist[orig] != 0) {
            if (otherend == 0) {
              i = orig;
              j = tail;
              k = m;
              l = trial;
              newpr = m;
            }
          }
          else {
            if (otherend != 0) {
              i = tail;
              j = orig;
              k = -m;
              l = -trial;
              newpr = m;
            }
          }
        }
      }
      trial++;
      lead = head[trial];
    } while ((lead ^ other) > 0);
    // prepare the next 'tail' group
    tail++;
    if (tail == nodes1) {
      tail = 1;
      trial = artedge1;
      lead = head[trial];
    }
    newprnb = false;
    if (newpr != maxint) {
      newprnb = true;
      break;
```

```
      }
    } while (tail != tailz);
    if (!newprnb) {
      for (p=1; p<=nodes; p++) {
        if (dist[p] != 0) continue;
        // add artificial
        sptpt[p] = artedge;
        flow[p] = 0;
        succ[p] = succ[nodes1];
        succ[nodes1] = p;
        pred[p] = nodes1;
        dist[p] = 1;
        dual[p] = -maxint;
      }
      break;
    }
    newprnb = false;
    sptpt[j] = l;
    pred[j] = i;
    succ[j] = succ[i];
    succ[i] = j;
    dist[j] = dist[i] + 1;
    dual[j] = dual[i] - k;
    newarc = Math.abs(l);
    room[newarc] = -room[newarc];
    treenodes++;
  } while (true);
  for (p=1; p<=nodes; p++) {
    q = Math.abs(sptpt[p]);
    room[q] = -room[q];
  }
  for (p=1; p<=sedge; p++)
    if (room[p] + maxint == 0)  room[p] = 0;
  room[artedge] = maxint;
  room[sedge] = maxint;

  // initialize price
  tail = 1;
  trial = artedge1;
  lead = head[trial];
  iterations = 0;

  // new iteration
  do {
    iterations++;
    // pricing basic edges
    tailz = tail;
    newpr = 0;
    do {
      ratez = -dual[tail];
```

```
      other = (lead >= 0 ? nodes1 : -nodes1);
      do {
        orig = Math.abs(lead);
        rate = dual[orig] + ratez - cost[trial];
        if (room[trial] < 0) rate = -rate;
        if (room[trial] != 0) {
          if (rate > newpr) {
            newarc = trial;
            newpr = rate;
            newtail = tail;
          }
        }
        trial++;
        lead = head[trial];
      } while ((lead ^ other) > 0);
      tail++;
      if (tail == nodes2) {
        tail = 1;
        trial = artedge1;
        lead = head[trial];
      }
      newprz = true;
      if (newpr != 0) {
        newprz = false;
        break;
      }
    } while (tail != tailz);
    if (newprz) {
      for (p=1; p<=edges; p++)
        flowsol[p] = 0;
      // prepare summary
      infeasible = false;
      value = valuez;
      for (p=1; p<=nodes; p++) {
        i = Math.abs(sptpt[p]);
        if ((flow[p] != 0) && (cost[i] == maxint)) infeasible = true;
        value += cost[i] * flow[p];
      }
      for (p=1; p<=lastslackedge; p++)
        if (room[p] < 0) {
        q = -room[p];
        value += cost[p] * q;
        }
      if (infeasible) return 4;
      arccost[0] = value;
      for (p=1; p<=nodes; p++) {
        q = Math.abs(sptpt[p]);
        room[q] = -flow[p];
      }
      solarc = 0;
```

```
      tail = 1;
      trial = artedge1;
      lead = head[trial];
      do {
        other = (lead >= 0 ? nodes1 : -nodes1);
        do {
          load = Math.max(0, -room[trial]) + least[trial];
          if (load != 0) {
            orig = Math.abs(lead);
            solarc++;
            arcsol[0][solarc] = orig;
            arcsol[1][solarc] = tail;
            flowsol[solarc] = load;
          }
          trial++;
          lead = head[trial];
        } while ((lead ^ other) > 0);
        tail++;
      } while (tail != nodes1);
      arcsol[0][0] = solarc;
      return 0;
    }

    // ration test
    newlead = Math.abs(head[newarc]);
    part = Math.abs(room[newarc]);
    jpart = 0;

    // cycle search
    ptr[2] = (room[newarc] >= 0 ? tedges1 : -tedges1);
    ptr[1] = -ptr[2];
    rim[1] = newlead;
    rim[2] = newtail;
    distdiff = dist[newlead] - dist[newtail];
    kpart = 1;
    if (distdiff < 0)  kpart = 2;
    if (distdiff != 0) {
      right = rim[kpart];
      point = ptr[kpart];
      q = Math.abs(distdiff);
      for (p=1; p<=q; p++) {
        if ((point ^ sptpt[right]) <= 0) {
          // increase flow
          i = Math.abs(sptpt[right]);
          spare = room[i] - flow[right];
          sparez = -right;
        }
        else {
          // decrease flow
          spare = flow[right];
```

```
        sparez = right;
      }
      if (part > spare) {
        part = spare;
        jpart = sparez;
        partz = false;
        if (part == 0) {
          partz = true;
          break;
        }
      }
      right = pred[right];
    }
    if (!partz) rim[kpart] = right;
  }
  if (!partz) {
    do {
      if (rim[1] ==  rim[2]) break;
      for (p=1; p<=2; p++) {
        right = rim[p];
        if ((ptr[p] ^ sptpt[right]) <= 0) {
          // increase flow
          i = Math.abs(sptpt[right]);
          spare = room[i] - flow[right];
          sparez = -right;
        }
        else {
          // decrease flow
          spare = flow[right];
          sparez = right;
        }
        if (part > spare) {
          part = spare;
          jpart = sparez;
          kpart = p;
          partz = false;
          if (part == 0) {
            partz = true;
            break;
          }
        }
        rim[p] = pred[right];
      }
    } while (true);
    if (!partz) left = rim[1];
  }
  partz = false;
  if (part != 0) {
    // update flows
    rim[1] = newlead;
```

```
        rim[2] = newtail;
        if (jpart != 0)  rim[kpart] = Math.abs(jpart);
        for (p=1; p<=2; p++) {
          right = rim[p];
          point = (ptr[p] >= 0 ? part : -part);
          do {
            if (right == left) break;
            flow[right] -= point * (sptpt[right] >= 0 ? 1 : -1);
            right = pred[right];
          } while (true);
        }
      }
      if (jpart == 0) {
        room[newarc] = -room[newarc];
        continue;
      }
      ipart = Math.abs(jpart);
      if (jpart <= 0) {
        j = Math.abs(sptpt[ipart]);
        // set old edge to upper bound
        room[j] = -room[j];
      }
      load = part;
      if (room[newarc] <= 0) {
        room[newarc] = -room[newarc];
        load = room[newarc] - load;
        newpr = -newpr;
      }
      if (kpart != 2) {
        vertex1 = newlead;
        vertex2 = newtail;
        curedge = -newarc;
        newpr = -newpr;
      }
      else {
        vertex1 = newtail;
        vertex2 = newlead;
        curedge = newarc;
      }

      // update tree
      i = vertex1;
      j = pred[i];
      distlen = dist[vertex2] + 1;
      if (part != 0) {
        point = (ptr[kpart]  >= 0 ? part: -part);
        do {
          // update dual variable
          dual[i] += newpr;
          n = flow[i];
```

```
        flow[i] = load;
        track = (sptpt[i] >= 0 ? 1 : -1);
        spointer = Math.abs(sptpt[i]);
        sptpt[i] = curedge;
        olddist = dist[i];
        distdiff = distlen - olddist;
        dist[i] = distlen;
        focal = i;
        do {
          after = succ[focal];
          if (dist[after] <= olddist) break;
          dist[after] += distdiff;
          dual[after] += newpr;
          focal = after;
        } while (true);
        k = j;
        do {
          l = succ[k];
          if (l == i) break;
          k = l;
        } while (true);
        ipartout = false;
        if (i == ipart) {
          ipartout = true;
          break;
        }
        load = n - point * track;
        curedge = -(track >= 0 ? spointer : -spointer);
        succ[k] = after;
        succ[focal] = j;
        k = i;
        i = j;
        j = pred[j];
        pred[i] = k;
        distlen++;
      } while (true);
    }
    if (!ipartout) {
      do {
        dual[i] += newpr;
        n = flow[i];
        flow[i] = load;
        track = (sptpt[i] >= 0 ? 1 : -1);
        spointer = Math.abs(sptpt[i]);
        sptpt[i] = curedge;
        olddist = dist[i];
        distdiff = distlen - olddist;
        dist[i] = distlen;
        focal = i;
        do {
```

```
        after = succ[focal];
        if (dist[after] <= olddist) break;
        dist[after] += distdiff;
        // udpate dual variable
        dual[after] += newpr;
        focal = after;
      } while (true);
      k = j;
      do {
        l = succ[k];
        if (l == i) break;
        k = l;
      } while (true);
      // test for leaving edge
      if (i == ipart) break;
      load = n;
      curedge = -(track >= 0 ? spointer : -spointer);
      succ[k] = after;
      succ[focal] = j;
      k = i;
      i = j;
      j = pred[j];
      pred[i] = k;
      distlen++;
    } while (true);
  }
  ipartout = false;
  succ[k] = after;
  succ[focal] = succ[vertex2];
  succ[vertex2] = vertex1;
  pred[vertex1] = vertex2;
} while (true);
}
```

Example:

In the following network, each edge is associated with three numbers: cost, lower bound, upper bound. Find the minimum cost network flow.

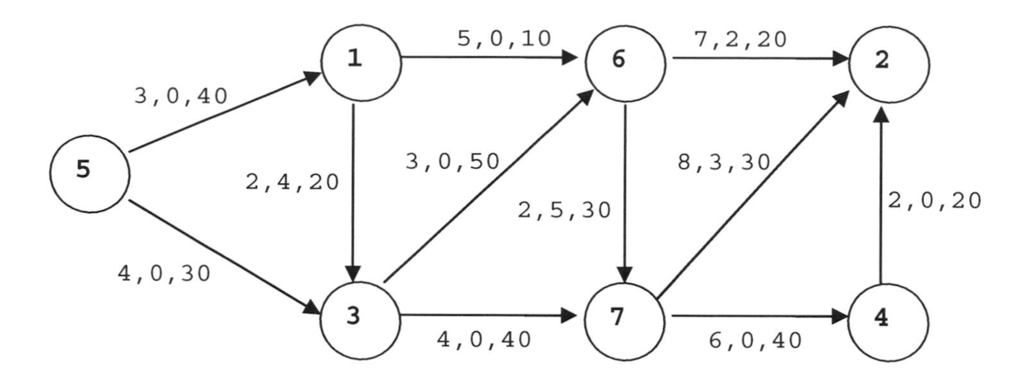


```
package GraphAlgorithms;

public class Test_mincostnetflow extends Object {

  public static void main(String args[]) {
    int nodes = 7;
    int edges = 11;
    int numdemand = 2;
    int nodedemand[][] = {{0, 0}, {5, 25}, {2, -25}};
    int nodei[]    = {0,  5,  1,  1,  7,  3,  4,  5,  6,  6,  7,  3};
    int nodej[]    = {0,  1,  3,  6,  4,  7,  2,  3,  2,  7,  2,  6};
    int arccost[]  = {0,  3,  2,  5,  6,  4,  2,  4,  7,  2,  8,  3};
    int lowbound[] = {0,  0,  4,  0,  0,  0,  0,  0,  2,  5,  3,  0};
    int upbound[]  = {0, 40, 20, 10, 40, 40, 20, 30, 20, 30, 30, 50};
    int f;
    int arcsol[][] = new int[2][edges + 1];
    int flowsol[] = new int[edges + 1];

    f = GraphAlgo.minCostNetworkFlow(nodes, edges, numdemand,
              nodedemand, nodei, nodej, arccost, upbound,
              lowbound, arcsol, flowsol);
    if (f > 0)
      System.out.println(" Infeasible, return code = " + f);
    else {
      System.out.println("Optimal solution found.\n  from to  flow");
      for (int i=1; i<=arcsol[0][0]; i++)
        System.out.printf("%5d %3d %4d\n",arcsol[0][i],arcsol[1][i],
                         flowsol[i]);
      System.out.println("\nTotal cost of the flow = " + arccost[0]);
    }
  }
}
```

Output:

```
Optimal solution found.
  from to  flow
    5   1    4
    4   2    2
    6   2   20
    7   2    3
    1   3    4
    5   3   21
    7   4    2
    3   6   25
    6   7    5

Total cost of the flow = 369
```

9. Packing and Covering

9.1 Assignment Problem

Suppose there are n persons to be assigned to n jobs. Given the cost c_{ij} of assigning the i-th person to the j-th job, the *assignment problem* is to assign exactly one job to each person with the minimum total cost. This can be formulated as an optimization problem:

$$\text{minimize} \quad \sum_{i=1}^{n} \sum_{j=1}^{n} c_{ij} x_{ij}$$

subject to

$$\sum_{i=1}^{n} x_{ij} = 1 \qquad (j = 1,2,\ldots,n)$$

$$\sum_{j=1}^{n} x_{ij} = 1 \qquad (i = 1,2,\ldots,n)$$

$$x_{ij} = 0 \text{ or } 1 \qquad (i = 1,2,\ldots,n, \quad j = 1,2,\ldots,n)$$

where x_{ij} has the value 1 if and only if person i is assigned to job j.

The Hungarian method [K55] is the classical method for solving the assignment problem. The following algorithm [CT80] improves on the Hungarian method by choosing a different initial solution and using a pointer implementation to locate the unexplored rows and zero elements of the cost matrix efficiently. At each iteration, an unassigned person is chosen and computes a shortest augmenting path from the person to the set of unassigned jobs using reduced costs as the edge lengths. The complementary slackness is maintained by adjusting the job prices. The process is repeated until all persons are assigned.

Procedure parameters:

void assignment (*n*, *cost*, *sol*)

n: int;
entry: *n* jobs to be assigned to *n* persons.

cost: int[*n*+2][*n*+2];
entry: *cost[i][j]*, is the cost of assigning person *i* to job *j*, *i*=1,2,...,*n*, *j*=1,2,...,*n*. The other elements of the matrix *cost* are not required as input.
exit: *cost[0][0]* is the total cost of the optimal solution. The cost matrix is modified on output.

sol: int[*n*+1];
exit: In the optimal solution, column *i* is assigned to row *sol[i]*, for i = 1, 2, ..., *n*.

```
public static void assignment(int n, int cost[][], int sol[])
{
  int h, i, j, k, tmpcol, tmprow, m, optcost, n1, temp;
  int tmpsetcol=0, tmpsetrow=0, r=0, l=0;
  int unexplorecol[] = new int[n + 1];
  int labelcol[] = new int[n + 1];
  int labelrow[] = new int[n + 1];
  int lastunassignrow[] = new int[n + 1];
  int nextunassignrow[] = new int[n + 1];
  int setlabelcol[] = new int[n + 1];
  int setlabelrow[] = new int[n + 1];
  int unexplorerow[] = new int[n + 2];
  int unassignrow[] = new int[n + 2];
  boolean skip=false, outer, middle;

  // initialize
  n1 = n + 1;
  for (j=1; j<=n; j++) {
    sol[j] = 0;
    lastunassignrow[j] = 0;
    nextunassignrow[j] = 0;
    unassignrow[j] = 0;
  }
  unassignrow[n1] = 0;
  optcost = 0;

  // cost matrix reduction
  for (j=1; j<=n; j++) {
    temp = cost[1][j];
    for (l=2; l<=n; l++)
      if (cost[l][j] < temp) temp = cost[l][j];
    optcost += temp;
    for (i=1; i<=n; i++)
      cost[i][j] -= temp;
  }
  for (i=1; i<=n; i++) {
    temp = cost[i][1];
    for (l=2; l<=n; l++)
      if (cost[i][l] < temp) temp = cost[i][l];
    optcost += temp;
    l = n1;
    for (j=1; j<=n; j++) {
      cost[i][j] -= temp;
      if ( cost[i][j] == 0 ) {
        cost[i][l] = -j;
        l = j;
      }
    }
  }
```

```
// choosing initial solution
k = n1;
for (i=1; i<=n; i++) {
  tmpcol = n1;
  j = -cost[i][n1];
  skip = false;
  do {
    if ( sol[j] == 0 ) {
      skip = true;
      break;
    }
    tmpcol = j;
    j = -cost[i][j];
  } while ( j != 0 );
  if (skip) {
    sol[j] = i;
    cost[i][tmpcol] = cost[i][j];
    nextunassignrow[i] = -cost[i][j];
    lastunassignrow[i] = tmpcol;
    cost[i][j] = 0;
    continue;
  }
  tmpcol = n1;
  j = -cost[i][n1];
  do {
    r = sol[j];
    tmprow = lastunassignrow[r];
    m = nextunassignrow[r];
    while (true) {
      if (m == 0) break;
      if (sol[m] == 0) {
        nextunassignrow[r] = -cost[r][m];
        lastunassignrow[r] = j;
        cost[r][tmprow] = -j;
        cost[r][j] = cost[r][m];
        cost[r][m] = 0;
        sol[m] = r;
        sol[j] = i;
        cost[i][tmpcol] = cost[i][j];
        nextunassignrow[i] = -cost[i][j];
        lastunassignrow[i] = tmpcol;
        cost[i][j] = 0;
        skip = true;
        break;
      }
      tmprow = m;
      m = -cost[r][m];
    }
    if (skip) break;
```

```
        tmpcol = j;
        j = -cost[i][j];
      } while ( j != 0 );
      if (skip) continue;
      unassignrow[k] = i;
      k = i;
    }
    middle = false;
    while (true) {
      outer = false;
      if (!middle) {
        if (unassignrow[n1] == 0) {
          cost[0][0] = optcost;
          return;
        }

        // search for a new assignment
        for (i=1; i<=n; i++) {
          unexplorecol[i] = 0;
          labelcol[i] = 0;
          labelrow[i] = 0;
          unexplorerow[i] = 0;
        }
        unexplorerow[n1] = -1;
        tmpsetcol = 0;
        tmpsetrow = 1;
        r = unassignrow[n1];
        labelrow[r] = -1;
        setlabelrow[1] = r;
      }
      if ((cost[r][n1] != 0) || middle) {
        do {
          if (!middle) {
            l = -cost[r][n1];
            if ( cost[r][l] != 0 ) {
              if ( unexplorerow[r] == 0 ) {
                unexplorerow[r] = unexplorerow[n1];
                unexplorecol[r] = -cost[r][l];
                unexplorerow[n1] = r;
              }
            }
            skip = false;
          }
          while (true) {
            if (!middle) {
              if (labelcol[l] == 0) break;
              if (unexplorerow[r] == 0) {
                skip = true;
                break;
              }
```

```
      }
      middle = false;
      l = unexplorecol[r];
      unexplorecol[r] = -cost[r][l];
      if ( cost[r][l] == 0 ) {
        unexplorerow[n1] = unexplorerow[r];
        unexplorerow[r] = 0;
      }
    }
    if (skip) break;
    labelcol[l] = r;
    if ( sol[l] == 0 ) {
      while (true) {
        // assigning a new row
        sol[l] = r;
        m = n1;
        while (true) {
          temp = -cost[r][m];
          if (temp == l) break;
          m = temp;
        }
        cost[r][m] = cost[r][l];
        cost[r][l] = 0;
        if (labelrow[r] < 0) break;
        l = labelrow[r];
        cost[r][l] = cost[r][n1];
        cost[r][n1] = -l;
        r = labelcol[l];
      }
      unassignrow[n1] = unassignrow[r];
      unassignrow[r] = 0;
      outer = true;
      break;
    }
    if (outer) break;
    tmpsetcol++;
    setlabelcol[tmpsetcol] = l;
    r = sol[l];
    labelrow[r] = l;
    tmpsetrow++;
    setlabelrow[tmpsetrow] = r;
  } while (cost[r][n1] != 0);
  if (outer) continue;
  if ( unexplorerow[n1] > 0 ) middle = true;
}
if (!middle) {
  // current cost matrix reduction
  h = Integer.MAX_VALUE;
  for (j=1; j<=n; j++)
    if ( labelcol[j] == 0 )
```

```
          for (k=1; k<=tmpsetrow; k++) {
            i = setlabelrow[k];
            if (cost[i][j] < h) h = cost[i][j];
          }
      optcost += h;
      for (j=1; j<=n; j++)
        if (labelcol[j] == 0)
          for (k=1; k<=tmpsetrow; k++) {
            i = setlabelrow[k];
            cost[i][j] -= h;
            if ( cost[i][j] == 0 ) {
              if ( unexplorerow[i] == 0 ) {
                unexplorerow[i] = unexplorerow[n1];
                unexplorecol[i] = j;
                unexplorerow[n1] = i;
              }
              l = n1;
              while (true) {
                temp = -cost[i][l];
                if (temp == 0) break;
                l = temp;
              }
              cost[i][l] = -j;
            }
          }
      if ( tmpsetcol != 0 )
        for (i=1; i<=n; i++) {
          if ( labelrow[i] == 0 ) {
            for (k=1; k<=tmpsetcol; k++) {
              j = setlabelcol[k];
              if ( cost[i][j] <= 0 ) {
                l = n1;
                while (true) {
                  temp = - cost[i][l];
                  if (temp == j) break;
                  l = temp;
                }
                cost[i][l] = cost[i][j];
                cost[i][j] = h;
                continue;
              }
              cost[i][j] += h;
            }
          }
      }
    }
    r = unexplorerow[n1];
    middle = true;
  }
}
```

Example:

Solve the assignment problem with the following cost matrix.

$$\text{Person}\ \overset{\text{Job}}{\begin{pmatrix} 9 & 6 & 8 & 4 & 7 \\ 8 & 7 & 6 & 8 & 3 \\ 7 & 4 & 3 & 5 & 6 \\ 9 & 8 & 3 & 7 & 6 \\ 8 & 7 & 3 & 6 & 5 \end{pmatrix}}$$

```
package GraphAlgorithms;

public class Test_assignment extends Object {

  public static void main(String args[]) {
    int n = 5;
    int cost[][] = {{0,  0,  0,  0,  0,  0,  0},
                    {0,  9,  6,  8,  4,  7,  0},
                    {0,  8,  7,  6,  8,  3,  0},
                    {0,  7,  4,  3,  5,  6,  0},
                    {0,  9,  8,  3,  7,  6,  0},
                    {0,  8,  7,  3,  6,  5,  0}};
    int sol[] = new int[n + 1];

    GraphAlgo.assignment(n, cost, sol);
    System.out.println("Optimal assignment:\n" + "  column  row");
    for (int i=1; i<=n; i++)
      System.out.println("     " + i + "  -  " + sol[i]);
    System.out.println("\nTotal assignment cost = " + cost[0][0]);
  }
}
```

Output:

```
Optimal assignment:
  column  row
     1  -  5
     2  -  3
     3  -  4
     4  -  1
     5  -  2

Total assignment cost = 22
```

9.2 Bottleneck Assignment Problem

Suppose there are n persons to be assigned to n jobs. Given the cost c_{ij} of assigning the i-th person to the j-th job, the *bottleneck assignment problem* is to minimize the maximum cost of assigning exactly one job to each person. This can be formulated as an optimization problem:

$$\text{minimize} \quad \max_{\substack{1\le i\le n \\ 1\le j\le n}} \{ c_{ij} : x_{ij} = 1 \}$$

subject to

$$\sum_{i=1}^{n} x_{ij} = 1 \qquad (j = 1,2,\ldots,n)$$

$$\sum_{j=1}^{n} x_{ij} = 1 \qquad (i = 1,2,\ldots,n)$$

$$x_{ij} = 0 \text{ or } 1 \qquad (i = 1,2,\ldots,n, \quad j = 1,2,\ldots,n)$$

where x_{ij} has the value 1 if and only if person i is assigned to job j.

The following procedure [JV87] solves the bottleneck assignment problem.

Procedure parameters:

void bottleneckAssignment (*n*, *cost*, *sol*)

n: int;
entry: the size of the $n \times n$ square matrix.

cost: int[*n*+1][*n*+1];
entry: *cost[i][j]*, is the cost of assigning person *i* to job *j*, *i*=1,2,...,*n*, *j*=1,2,...,*n*.
exit: *cost[0][0]* is the bottleneck value in the optimal solution. The cost matrix is not modified on output.

sol: int[*n*+1];
exit: In the optimal solution, row *i* is assigned to column *sol[i]*, for i = 1, 2, ..., *n*.

```
public static void bottleneckAssignment(int n, int cost[][], int sol[])
{
  int i, minimum, bottleneckvalue, cand, p, v, w, start, temp;
  int q=0, u=0;
  int aux1[] = new int[n + 1];
  int aux2[] = new int[n + 1];
  int aux3[] = new int[n + 1];
  int aux4[] = new int[n + 1];
  boolean skip=false;
```

```
// initialize
bottleneckvalue = -Integer.MAX_VALUE;
for (i=1; i<=n; i++)
  aux1[i]=0;
for (i=0; i<=n-1; i++) {
  q = n - i;
  aux2[q] = q;
  minimum = Integer.MAX_VALUE;
  p = 1;
  do {
    cand = cost[q][p];
    if (cand <= bottleneckvalue) {
      minimum = bottleneckvalue;
      u = p;
      p = n + 1;
    }
    else {
      if (cand < minimum) {
        minimum = cand;
        u = p;
      }
      p++;
    }
  } while (p <= n);
  if (aux1[u] == 0) {
    aux1[u] = q;
    sol[q] = u;
  }
  else
    sol[q] = 0;
  if (minimum > bottleneckvalue) bottleneckvalue = minimum;
}
// search for an augmenting path
for (u=1; u<=n; u++)
  if (aux1[u] == 0) {
    w = 1;
    start = 1;
    skip = false;
    for (i=1; i<=n; i++) {
      q = aux2[i];
      aux4[q] = cost[q][u];
      aux3[q] = u;
      if (aux4[q] <= bottleneckvalue) {
        if (sol[q] == 0) {
          skip = true;
          break;
        }
        aux2[i] = aux2[w];
        aux2[w] = q;
        w++;
```

```
    }
  }
  if (!skip) {
    while (true) {
      if (w == start) {
        minimum = Integer.MAX_VALUE;
        for (i=w; i<=n; i++) {
          q = aux2[i];
          cand = aux4[q];
          if (cand <= minimum) {
            if (cand < minimum) {
              w = start;
              minimum = cand;
            }
            aux2[i] = aux2[w];
            aux2[w] = q;
            w++;
          }
        }
        bottleneckvalue = minimum;
        for (i=start; i<=w-1; i++) {
          q = aux2[i];
          if (sol[q] == 0) {
            skip = true;
            break;
          }
        }
      }
      if (skip) break;
      v = aux2[start];
      start++;
      p = sol[v];
      for (i=w; i<=n; i++) {
        q = aux2[i];
        cand = cost[q][p];
        if (cand < aux4[q]) {
          aux3[q] = p;
          if (cand <= bottleneckvalue) {
            if (sol[q] == 0) {
              skip = true;
              break;
            }
            aux2[i] = aux2[w];
            aux2[w] = q;
            w++;
          }
          aux4[q] = cand;
        }
      }
      if (skip) break;
```

```
        }
      }
      // augment
      skip = false;
      do {
        p = aux3[q];
        sol[q] = p;
        temp = q;
        q = aux1[p];
        aux1[p] = temp;
      } while (p != u);
    }
  cost[0][0] = bottleneckvalue;
}
```

Example:

Solve the bottleneck assignment problem with the following cost matrix.

$$\text{Person}\ \overset{\text{Job}}{\begin{pmatrix} 9 & 6 & 8 & 4 & 7 \\ 8 & 7 & 6 & 8 & 3 \\ 7 & 4 & 3 & 5 & 6 \\ 9 & 8 & 3 & 7 & 6 \\ 8 & 7 & 3 & 6 & 5 \end{pmatrix}}$$

```
package GraphAlgorithms;

public class Test_bottleneckAssignment extends Object {

  public static void main(String args[]) {

    int n = 5;
    int cost[][] = {{0,  0,  0,  0,  0,  0,  0},
                    {0,  9,  6,  8,  4,  7,  0},
                    {0,  8,  7,  6,  8,  3,  0},
                    {0,  7,  4,  3,  5,  6,  0},
                    {0,  9,  8,  3,  7,  6,  0},
                    {0,  8,  7,  3,  6,  5,  0}};
    int sol[] = new int[n + 1];

    GraphAlgo.bottleneckAssignment(n, cost, sol);
    System.out.println("Optimal solution:\n" + "     row    column");
    for (int i=1; i<=n; i++)
      System.out.println("      " + i + "   -   " + sol[i]);
    System.out.println("\nBottleneck value = " + cost[0][0]);
  }
}
```

Output:

```
Optimal solution:
    row   column
     1  -  2
     2  -  5
     3  -  1
     4  -  4
     5  -  3

Bottleneck value = 7
```

9.3 Quadratic Assignment Problem

Suppose n facilities are to be established in n locations. The nonnegative distance a_{ij} between any two locations is known, and the nonnegative weight b_{ij} between any two facilities is given, i=1,2,...,n, j=1,2,...,n. The *quadratic assignment problem* is to assign all facilities to different locations so that the sum of the distances multiplied by the corresponding weights is minimized. The problem can be formulated as follows:

$$\text{minimize} \quad \sum_{i=1}^{n}\sum_{j=1}^{n}\sum_{u=1}^{n}\sum_{v=1}^{n} a_{iu} b_{jv} x_{ij} x_{uv}$$

subject to

$$\sum_{i=1}^{n} x_{ij} = 1 \qquad (j = 1,2,\ldots,n)$$

$$\sum_{j=1}^{n} x_{ij} = 1 \qquad (i = 1,2,\ldots,n)$$

$$x_{ij} = 0 \text{ or } 1 \qquad (i = 1,2,\ldots,n, \quad j = 1,2,\ldots,n)$$

x_{ij} has the value 1 if and only if i is assigned to j.

The following procedure [BD80, PC71] solves this NP-hard problem by implicit enumeration.

Procedure parameters:

void quadraticAssignment (*n*, *a*, *b*, *sol*)

n: int;
entry: *n* facilities to be assigned to *n* locations.

a: int[*n*+1][*n*+1];
entry: *a[i][j]*, is the nonnegative distance between location i and location j, *i*=1,2,...,*n*, *j*=1,2,...,*n*. The matrix *a* is assumed to be symmetric.
exit: *cost[0][0]* is the total cost of the optimal solution. The cost matrix is modified on output.

b: int[*n*+1][*n*+1];
entry: *b[i][j]*, is the nonnegative weight between facility i and facility j, *i*=1,2,...,*n*, *j*=1,2,...,*n*. The matrix *b* is assumed to be symmetric.

sol: int[*n*+1];
exit: In the optimal solution, *i* is assigned to *sol[i]*, for i = 1, 2, ..., *n*.
sol[0] is the optimal total cost of the objective function.

```
public static void quadraticAssignment(int n, int a[][], int b[][], int sol[])
{
  int valc,vald,valf,valg,valh,vali,valk,valp,valq,valw,valx,valy;
  int i,j,k,p,rowdist,rowweight,rowmata,rowmatb,partialobj;
  int leastw,leastx,leasty,matbx,matby,matcx,matcy,matax;
  int bestobjval,ntwo,npp,large;
  int vale=0,subcripta=0,subcriptb=0,subcriptc=0,subcriptd=0,partialz=0;
  int parm1[] = new int[1];
  int parm2[] = new int[1];
  int parm3[] = new int[1];
  int aux2[] = new int[n-1];
  int aux3[] = new int[n-1];
  int aux4[] = new int[n-1];
  int aux5[] = new int[n-1];
  int aux6[] = new int[n-1];
  int aux7[] = new int[n+1];
  int aux8[] = new int[n+1];
  int aux9[] = new int[n+1];
  int aux10[] = new int[n+1];
  int aux11[] = new int[n+1];
  int aux12[] = new int[n+1];
  int aux13[] = new int[n+1];
  int aux14[] = new int[n+1];
  int aux15[] = new int[n*n+1];
  int aux16[] = new int[n*(n+1)*(2*n-2)/6 + 1];
  int aux17[] = new int[n*(n+1)*(2*n-2)/6 + 1];
  int aux18[] = new int[(n*(n+1)*(2*n+1)/6)];
  int aux19[][] = new int[n+1][n+1];
  int aux20[][] = new int[n+1][n+1];
  boolean work1[] = new boolean[n+1];
  boolean work2[] = new boolean[n+1];
  boolean work3[] = new boolean[n+1];
  boolean contr,skip=false;

  // initialization
  p = 1;
  k = 1;
```

```
for (i=1; i<=n; i++) {
  for (j=1; j<=n; j++) {
    p += a[i][j];
    k += b[i][j];
    aux19[i][j] = 0;
  }
  aux13[i] = 0;
}
large = n * p * k;
k = 0;
bestobjval = large;
ntwo = n - 2;
valy = n + 1;
j = 0;
valp = 0;
for (i=1; i<=ntwo; i++) {
  npp = valy - i;
  valq = npp * npp;
  j += valq - npp;
  aux4[i] = j;
  valp += valq;
  aux5[i] = valp;
}
// compute a[i][i] * b[j][j]
for (i=1; i<=n; i++) {
  work1[i] = false;
  work2[i] = false;
  partialz = a[i][i];
  for (j=1; j<=n; j++) {
    aux20[i][j] = -1;
    aux19[i][j] += partialz * b[j][j];
  }
}
for (j=1; j<=n; j++) {
  a[j][j] = large;
  b[j][j] = large;
}
// reduce matrices a and b
for (i=1; i<=n; i++) {
  rowmata = a[i][1];
  rowmatb = b[i][1];
  for (j=2; j<=n; j++) {
    rowdist = a[i][j];
    rowweight = b[i][j];
    if (rowdist < rowmata) rowmata = rowdist;
    if (rowweight < rowmatb) rowmatb = rowweight;
  }
  for (j=1; j<=n; j++) {
    a[i][j] -= rowmata;
    b[i][j] -= rowmatb;
```

```
    }
    aux7[i] = rowmata;
    aux8[i] = rowmatb;
  }
  // rowwise reduction of matrices a and b
  for (i=1; i<=n; i++) {
    matcy = aux7[i];
    for (j=1; j<=n; j++) {
      matbx = aux8[j];
      matby = (n-1) * matbx;
      for (p=1; p<=n; p++)
        if (p != j) matby += b[j][p];
      matcx = matcy * matby;
      matax = 0;
      for (p=1; p<=n; p++)
        if (p != i) matax += a[i][p];
      aux19[i][j] += matcx + matbx * matax;
    }
  }
  // columnwise reduction of matrices a and b
  for (i=1; i<=n; i++) {
    rowmata = a[1][i];
    rowmatb = b[1][i];
    for (j=2; j<=n; j++) {
      rowdist = a[j][i];
      rowweight = b[j][i];
      if (rowdist < rowmata) rowmata = rowdist;
      if (rowweight < rowmatb) rowmatb = rowweight;
    }
     for (j=1; j<=n; j++) {
       a[j][i] -= rowmata;
       b[j][i] -= rowmatb;
    }
    aux7[i] = rowmata;
    aux8[i] = rowmatb;
  }
  for (i=1; i<=n; i++) {
    a[i][i] = 0;
    b[i][i] = 0;
    matcy = aux7[i];
    for (j=1; j<=n; j++) {
      matbx = aux8[j];
      matby = (n-1) * matbx;
      for (p=1; p<=n; p++)
        if (p != j) matby += b[p][j];
      matcx = matcy * matby;
      matax = 0;
      for (p=1; p<=n; p++)
        if (p != i) matax += a[p][i];
      aux19[i][j] += matcx + matbx * matax;
```

```
    }
  }
  partialobj = 0;
  npp = n;
  contr = true;
  // compute minimal scalar products
  qapsubprog2(n,k,npp,subcripta,subcriptb,a,b,aux4,aux7,
              work1,work2,aux16,aux17,aux13);
  qapsubprog3(n,k,npp,aux15,aux4,aux5,contr,aux16,aux17,aux18);
  contr = false;
  iterate:
  while (true) {
    valx = 0;
    valy = 0;
    valq = 0;
    valf = 0;
    for (i=1; i<=n; i++)
      if (!work1[i]) {
        valc = valx * npp;
        valx++;
        vald = 0;
        for (j=1; j<=n; j++)
          if (!work2[j]) {
            vald++;
            valc++;
            if (aux20[i][j] < 0)
              aux15[valc] +- aux19[i][j];
            else {
              aux15[valc] = large;
              valf++;
              if (valf < 2) {
                subcriptc = valx;
                subcriptd = vald;
              }
              else {
                if (valf == 2) {
                  if (valx == subcriptc) valy = valx;
                  if (vald == subcriptd) valq = vald;
                }
              }
            }
          }
      }
    // obtain a bound by solving the linear assignment problem
    skip = false;
    obtainbound:
    while (true) {
      if (!skip) {
        qapsubprog4(npp,large,aux15,parm1,aux14,aux12,aux11,
                    aux9,aux7,aux8,aux13,work3);
```

```
    partialz = parm1[0];
    valc = 0;
    for (i=1; i<=npp; i++)
      for (j=1; j<=npp; j++) {
        valc++;
        aux15[valc] -= (aux7[i] + aux8[j]);
      }
    if (partialobj + partialz >= bestobjval) {
      // backtrack
      if (!contr) {
        if (k == 0) {
          return;
        }
        subcripta = aux2[k];
        subcriptb = aux3[k];
      }
      else {
        contr = false;
        k++;
        // cancel the last single assignment
        for (i=1; i<=n; i++)
          if (!work1[i]) {
            for (j=1; j<=n; j++)
              if (!work2[j] && aux20[i][j] == k) aux20[i][j] = -1;
          }
        partialobj -= aux19[subcripta][subcriptb];
        work1[subcripta] = false;
        work2[subcriptb] = false;
        k--;
        npp = n - k;
        if (aux6[k+1] + partialobj >= bestobjval) {
          if (k == 0) {
            return;
          }
          subcripta = aux2[k];
          subcriptb = aux3[k];
        }
        else {
          qapsubprog3(n,k,npp,aux15,aux4,aux5,contr,aux16,aux17,aux18);
          continue iterate;
        }
      }
      skip = true;
      continue obtainbound;
    }
    if (contr) {
      skip = true;
      continue obtainbound;
    }
    skip = false;
```

```
    }
    if (skip) {
      skip = false;
      // the solution tree is completed
      for (i=1; i<=n; i++)
        if (!work1[i]) {
          valh = a[subcripta][i];
          vali = a[i][subcripta];
          for (j=1; j<=n; j++)
            if (!work2[j]) {
              valg = valh * b[subcriptb][j] + vali * b[j][subcriptb];
              if (!contr) valg = -valg;
              aux19[i][j] += valg;
            }
        }
      if (!contr) {
        // cancel the last single assignment
        for (i=1; i<=n; i++)
          if (!work1[i]) {
            for (j=1; j<=n; j++)
              if (!work2[j] && aux20[i][j] == k) aux20[i][j] = -1;
          }
        partialobj -= aux19[subcripta][subcriptb];
        work1[subcripta] = false;
        work2[subcriptb] = false;
        k--;
        npp = n - k;
        if (aux6[k+1] + partialobj >= bestobjval) {
          if (k == 0) {
            return;
          }
          subcripta = aux2[k];
          subcriptb = aux3[k];
          skip = true;
          continue obtainbound;
        }
        qapsubprog3(n,k,npp,aux15,aux4,aux5,contr,aux16,aux17,aux18);
        continue iterate;
      }
      aux10[subcripta] = subcriptb;
      k++;
      aux2[k] = subcripta;
      aux3[k] = subcriptb;
      if (k == (n-2)) {
        // compute the objective function values
        for (i=1; i<=n; i++)
          if (!work1[i]) {
            valx = i;
            break;
          }
```

```
        for (i=1; i<=n; i++)
          if (!work2[i]) {
            j = i;
            break;
          }
        work1[valx] = true;
        work2[j] = true;
        for (i=1; i<=n; i++)
          if (!work1[i]) {
            vale = i;
            break;
          }
        for (i=1; i<=n; i++)
          if (!work2[i]) {
            valp = i;
            break;
          }
        contr = false;
        valw = 0;
        while (true) {
          partialz = aux19[valx][j] + aux19[vale][valp] +
             a[valx][vale] * b[j][valp] + a[vale][valx] * b[valp][j];
          work1[valx] = false;
          work2[j] = false;
          if ((partialz + partialobj) < bestobjval) {
            bestobjval = partialz + partialobj;
            sol[0] = bestobjval;
            for (i=1; i<=n; i++)
              if (work1[i]) sol[i] = aux10[i];
            sol[valx] = j;
            sol[vale] = valp;
          }
          if (valw != 0) {
            if (k == 0) {
              return;
            }
            subcripta = aux2[k];
            subcriptb = aux3[k];
            skip = true;
            continue obtainbound;
          }
          valw = valx;
          valx = vale;
          vale = valw;
        }
      }
      valf = 0;
    }
    if (valf < 1) {
      // compute the alternative costs
```

```
    qapsubprog1(npp,aux15,aux12,large,parm1,parm2,parm3);
    subcripta = parm1[0];
    subcriptb = parm2[0];
    valk = parm3[0];
  }
  else {
    if (valf == 1) {
      // compute the next single assignment
      leastw = large;
      valp = aux12[subcriptc];
      valc = (subcriptc - 1) * npp;
      for (j=1; j<=npp; j++) {
        valc++;
        valg = aux15[valc];
        if ((leastw > valg) && (j != valp)) leastw = valg;
      }
      leastx = leastw;
      leastw = large;
      valc = valp;
      for (i=1; i<=npp; i++) {
        valg = aux15[valc];
        valc += npp;
        if ((leastw > valg) && (i != subcriptc)) leastw = valg;
      }
      leastx += leastw;
      leastw = large;
      valc = subcriptd;
      for (i=1; i<=npp; i++) {
        valg = aux15[valc];
        valc += npp;
        if ((valg < leastw) && (subcriptd != aux12[i])) leastw = valg;
      }
      leasty = leastw;
      i = 1;
      while (i <= npp) {
        if (aux12[i] == subcriptd) break;
        i++;
      }
      vale = i;
      valc = (vale - 1) * npp;
      leastw = large;
      for (j=1; j<=npp; j++) {
        valc++;
        valg = aux15[valc];
        if ((valg < leastw) && (j != subcriptd)) leastw = valg;
      }
      if ((leastw + leasty) >= leastx) {
        subcripta = vale;
        subcriptb = subcriptd;
        valk = leastw + leasty;
```

```
      }
      else {
        subcripta = subcriptc;
        subcriptb = valp;
        valk = leastx;
      }
    }
    else {
      // compute the next single assignment
      if (valy != 0) {
        subcripta = valy;
        subcriptb = aux12[subcripta];
      }
      else {
        subcriptb = valq;
        i = 1;
        while (i <= npp) {
          if (aux12[i] == subcriptb) break;
          i++;
        }
        subcripta = i;
      }
      leastw = large;
      valc = (subcripta - 1) * npp;
      for (i=1; i<=npp; i++) {
        valc++;
        valg = aux15[valc];
        if ((valg < leastw) && (i != subcriptb)) leastw = valg;
      }
      valk = leastw;
      leastw = large;
      valc = subcriptb;
      for (j=1; j<=npp; j++) {
        valg = aux15[valc];
        valc += npp;
        if ((valg < leastw) && (j != subcripta)) leastw = valg;
      }
      valk += leastw;
    }
  }
  valx = 0;
  aux6[k+1] = valk + partialz;
  i = 1;
  while (i <= n) {
    if (!work1[i]) {
      valx++;
      if (subcripta == valx) break;
    }
    i++;
  }
```

```
        subcripta = i;
        valx = 0;
        j = 1;
        while (j <= n) {
          if (!work2[j]) {
            valx++;
            if (subcriptb == valx) break;
          }
          j++;
        }
        subcriptb = j;
        aux20[subcripta][subcriptb] = k;
        contr = true;
        work1[subcripta] = true;
        work2[subcriptb] = true;
        npp = n - k - 1;
        // compute the cost matrix
        qapsubprog2(n,k,npp,subcripta,subcriptb,a,b,aux4,aux7,
                    work1,work2,aux16,aux17,aux13);
        qapsubprog3(n,k,npp,aux15,aux4,aux5,contr,aux16,aux17,aux18);
        valx = 0;
        for (i=1; i<=n; i++)
          if (!work1[i]) {
            valh - a[i][subcripta];
            vali = a[subcripta][i];
            vald = valx;
            for (j=1; j<-n; j++)
              if (!work2[j]) {
                vald++;
                aux15[vald] += aux19[i][j] + valh * b[j][subcriptb] +
                                             vali * b[subcriptb][j];
              }
            valx += npp;
          }
        partialobj += aux19[subcripta][subcriptb];
      }
    }
  }


static private void qapsubprog1(int n, int aux15[], int aux12[],
               int large, int parm1[], int parm2[], int parm3[])
{
  /* this method is used internally by quadraticAssignment */

  // compute the alternative costs and obtain the assignment

  int i,j,leastw,leastx,p,q,valc,valp;

  parm3[0] = -1;
```

```
  valc = 0;
  for (i=1; i<=n; i++) {
    j = aux12[i];
    valp = j - n;
    leastw = large;
    for (p=1; p<=n; p++) {
      valp += n;
      if (p != i) {
        q = aux15[valp];
        if (q < leastw) leastw = q;
      }
    }
    leastx = leastw;
    leastw = large;
    for (p=1; p<=n; p++) {
      valc++;
      if (p != j) {
        q = aux15[valc];
        if (q < leastw) leastw = q;
      }
    }
    leastw += leastx;
    if (leastw > parm3[0]) {
      parm1[0] = i;
      parm2[0] = j;
      parm3[0] = leastw;
    }
  }
}


static private void qapsubprog2(int n, int k, int npp, int subcripta,
                     int subcriptb, int a[][], int b[][], int aux4[],
                     int vekt[], boolean work1[], boolean work2[],
                     int aux16[], int aux17[], int aux13[])
{
  /* this method is used internally by quadraticAssignment */

  // obtain rows of the matrix a in decreasing order
  // and rows of matrix b in increasing order

  int i,j,nppa,nppb,nppc,valg,valp,valq,valx;
  boolean decide;

  nppa = npp - 1;
  if (npp == n) {
    valp = 1;
    for (i=1; i<=n; i++) {
      valx = 0;
      for (j=1; j<=n; j++)
```

```
        if (j != i) {
          valx++;
          vekt[valx] = a[i][j];
        }
      qapsubprog5(vekt,aux13,nppa);
      for (j=1; j<=nppa; j++) {
        nppb = npp - j;
        aux16[valp] = vekt[nppb];
        valp++;
      }
    }
    valp = 1;
    for (i=1; i<=n; i++) {
      valx = 0;
      for (j=1; j<=n; j++)
        if (j != i) {
          valx++;
          vekt[valx] = b[i][j];
        }
      qapsubprog5(vekt,aux13,nppa);
      for (j=1; j<=nppa; j++) {
        aux17[valp] = vekt[j];
        valp++;
      }
    }
    return;
  }
  nppc = aux4[k+1];
  valp = nppc;
  valq = nppc - (npp + 1) * npp;
  for (i=1; i<=n; i++)
    if (work1[i]) {
      if (i == subcripta) valq += npp;
    }
    else {
      valg = a[i][subcripta];
      decide = true;
      for (j=1; j<=npp; j++) {
        valq++;
        if ((aux16[valq] == valg) && decide)
          decide = false;
        else {
          valp++;
          aux16[valp] = aux16[valq];
        }
      }
    }
  valp = nppc;
  valq = nppc - (npp + 1) * npp;
  for (i=1; i<=n; i++) {
```

```
    if (work2[i]) {
      if (i == subscriptb) valq += npp;
    }
    else {
      valg = b[i][subscriptb];
      decide = true;
      for (j=1; j<=npp; j++) {
        valq++;
        if ((aux17[valq] == valg) && decide)
          decide = false;
        else {
          valp++;
          aux17[valp] = aux17[valq];
        }
      }
    }
  }
  return;
}


static private void qapsubprog3 (int n, int k, int npp, int aux15[],
                 int aux4[], int aux5[], boolean contr, int aux16[],
                 int aux17[], int aux18[])
{
  /* this method is used internally by quadraticAssignment */

  // compute the cost matrix of the k-th linear assignment problem

  int accum,i,j,idx,nppd,valg,valp,valq,valr,vals,valt,valx;

  idx = 0;
  if (npp == n) {
    accum = 0;
    valp = 0;
    if (contr) {
      valq = accum;
      nppd = npp - 1;
      for (i=1; i<=npp; i++) {
        valt = accum;
        for (j=1; j<=npp; j++) {
          valp++;
          idx++;
          valg = 0;
          for (valx=1; valx<=nppd; valx++) {
            valr = valq + valx;
            vals = valt + valx;
            valg += aux16[valr] * aux17[vals];
          }
          valt += nppd;
```

```
          aux18[valp] = valg;
          aux15[idx] = valg;
        }
        valq += nppd;
      }
      return;
    }
    for (i=1; i<=npp; i++)
      for (j=1; j<=npp; j++) {
        valp++;
        idx++;
        aux15[idx] = aux18[valp];
      }
    return;
  }
  if (!contr) {
    valp = aux5[k];
    for (i=1; i<=npp; i++)
      for (j=1; j<=npp; j++) {
        valp++;
        idx++;
        aux15[idx] = aux18[valp];
      }
    return;
  }
  accum = aux4[k+1];
  valp = aux5[k+1];
  valq = accum;
  nppd = npp - 1;
  for (i=1; i<=npp; i++) {
    valt = accum;
    for (j=1; j<=npp; j++) {
      valp++;
      idx++;
      valg = 0;
      for (valx=1; valx<=nppd; valx++) {
        valr = valq + valx;
        vals = valt + valx;
        valg += aux16[valr] * aux17[vals];
      }
      valt += nppd;
      aux18[valp] = valg;
      aux15[idx] = valg;
    }
    valq += nppd;
  }
  return;
}
```

```
static private void qapsubprog4(int n, int large, int c[], int parm[],
                int wk1[], int wk2[], int wk3[], int wk4[], int wk5[],
                int wk6[], int wk7[], boolean wk8[])
{
  /* this method is used internally by quadraticAssignment */

  // solve linear sum assignment problem

  int i,j,p,v1,v2,v5,v6,v7,v8,v9,v10,v11,v12,v13,v14,v15;
  int v3=0,v4=0,v16=0;

  // initial assignment
  for (i=1; i<=n; i++) {
    wk1[i] = 0;
    wk2[i] = 0;
    wk5[i] = 0;
    wk6[i] = 0;
    wk7[i] = 0;
  }
  v1=0;
  for (i=1; i<=n; i++) {
    for (j=1; j<=n; j++) {
      v1++;
      v2 = c[v1];
      if (j == 1) {
        v3 = v2;
        v4 = j;
      }
      else {
        if ((v2 - v3) < 0) {
          v3 = v2;
          v4 = j;
        }
      }
    }
    wk5[i] = v3;
    if (wk1[v4] == 0) {
      wk1[v4] = i;
      wk2[i] = v4;
    }
  }
  for (j=1; j<=n; j++) {
    wk6[j] = 0;
    if (wk1[j] == 0) wk6[j] = large;
  }
  v1 = 0;
  for (i=1; i<=n; i++) {
    v3 = wk5[i];
    for (j=1; j<=n; j++) {
      v1++;
```

```
        v15 = wk6[j];
        if (v15 > 0) {
          v2 = c[v1] - v3;
          if (v2 < v15) {
            wk6[j] = v2;
            wk7[j] = i;
          }
        }
      }
    }
    for (j=1; j<=n; j++) {
      i = wk7[j];
      if (i != 0) {
        if (wk2[i] == 0) {
          wk2[i] = j;
          wk1[j] = i;
        }
      }
    }
    for (i=1; i<=n; i++)
      if (wk2[i] == 0) {
        v3 = wk5[i];
        v1 = (i - 1) * n;
        for (j=1; j<=n; j++) {
          v1++;
          if (wk1[j] == 0) {
            v2 = c[v1];
            if ((v2 - v3 - wk6[j]) <= 0) {
              wk2[i] = j;
              wk1[j] = i;
              break;
            }
          }
        }
      }
    // construct the optimal assignment
    for (p=1; p<=n; p++)
      if (wk2[p] <= 0) {
        // compute shortest path
        v5 = (p - 1) * n;
        for (i=1; i<=n; i++) {
          wk7[i] = p;
          wk8[i] = false;
          wk4[i] = large;
          v6 = v5 + i;
          wk3[i] = c[v6] - wk5[p] - wk6[i];
        }
        wk4[p] = 0;
        while (true) {
          v14 = large;
```

```
        for (i=1; i<=n; i++)
          if (!wk8[i]) {
            if (wk3[i] < v14) {
              v14 = wk3[i];
              v16 = i;
            }
          }
        if (wk1[v16] <= 0) break;
        wk8[v16] = true;
        v7 = wk1[v16];
        v8 = (v7 - 1) * n;
        wk4[v7] = v14;
        for (i=1; i<=n; i++)
          if (!wk8[i]) {
            v9 = v8 + i;
            v10 = v14 + c[v9] - wk5[v7] - wk6[i];
            if (wk3[i] > v10) {
              wk3[i] = v10;
              wk7[i] = v7;
            }
          }
      }
      // augmentation
      while (true) {
        v7 = wk7[v16];
        wk1[v16] = v7;
        v11 = wk2[v7];
        wk2[v7] = v16;
        if (v7 == p) break;
        v16 = v11;
      }
      // transformation
      for (i=1; i<=n; i++) {
        if (wk4[i] != large)
          wk5[i] += v14 - wk4[i];
        if (wk3[i] < v14)
          wk6[i] += wk3[i] - v14;
      }
    }
  // compute the optimal value
  parm[0] = 0;
  for (i=1; i<=n; i++) {
    v12 = (i - 1) * n;
    j = wk2[i];
    v13 = v12 + j;
    parm[0] += c[v13];
  }
}
```

```
static private void qapsubprog5 (int a[], int b[], int dim)
{
  /* this method is used internally by quadraticAssignment */

  // sort the vector a in increasing order
  // b is the permutation vector of the sorted vector

  int i,j,ina,inb,inx,iny,low,half,high,p,quant;

  low = 1;
  if (dim <= low) return;
  half = (dim - low + 1) / 2;
  quant = 1023;
  for (p=1; p<=10; p++) {
    if (quant <= half) {
      high = dim - quant;
      for (i=low; i<=high; i++) {
        inx = i + quant;
        ina = a[inx];
        inb = b[inx];
        j = i;
        iny = inx;
        while (ina < a[j]) {
          a[iny] = a[j];
          b[iny] = b[j];
          iny = j;
          j -= quant;
          if (j < low) break;
        }
        a[iny] = ina;
        b[iny] = inb;
      }
    }
    quant /= 2;
  }
}
```

Example:

Solve the quadratic assignment problem with the following given distance matrix and weight matrix.

n = 7,

$$\text{Distance matrix} = \begin{pmatrix} 0 & 3 & 5 & 1 & 6 & 4 & 1 \\ 3 & 0 & 5 & 1 & 9 & 2 & 8 \\ 5 & 5 & 0 & 2 & 4 & 5 & 1 \\ 1 & 1 & 2 & 0 & 5 & 9 & 1 \\ 6 & 9 & 4 & 5 & 0 & 2 & 3 \\ 4 & 2 & 5 & 9 & 2 & 0 & 6 \\ 1 & 8 & 1 & 1 & 3 & 6 & 0 \end{pmatrix}$$

$$\text{Weight matrix} = \begin{pmatrix} 0 & 2 & 6 & 1 & 3 & 0 & 5 \\ 2 & 0 & 5 & 8 & 0 & 8 & 1 \\ 6 & 5 & 0 & 0 & 5 & 3 & 4 \\ 1 & 8 & 0 & 0 & 4 & 3 & 8 \\ 3 & 0 & 5 & 4 & 0 & 3 & 0 \\ 0 & 8 & 3 & 3 & 3 & 0 & 2 \\ 5 & 1 & 4 & 8 & 0 & 2 & 0 \end{pmatrix}$$

```
package GraphAlgorithms;

public class Test_quadraticAssignment extends Object {

  public static void main(String args[]) {

    int n=7;
    int a[][] = {{0, 0, 0, 0, 0, 0, 0, 0},
                 {0, 0, 3, 5, 1, 6, 4, 1},
                 {0, 3, 0, 5, 1, 9, 2, 8},
                 {0, 5, 5, 0, 2, 4, 5, 1},
                 {0, 1, 1, 2, 0, 5, 9, 1},
                 {0, 6, 9, 4, 5, 0, 2, 3},
                 {0, 4, 2, 5, 9, 2, 0, 6},
                 {0, 1, 8, 1, 1, 3, 6, 0}};
```

```
    int b[][] = {{0, 0, 0, 0, 0, 0, 0, 0},
                 {0, 0, 2, 6, 1, 3, 0, 5},
                 {0, 2, 0, 5, 8, 0, 8, 1},
                 {0, 6, 5, 0, 0, 5, 3, 4},
                 {0, 1, 8, 0, 0, 4, 3, 8},
                 {0, 3, 0, 5, 4, 0, 3, 0},
                 {0, 0, 8, 3, 3, 3, 0, 2},
                 {0, 5, 1, 4, 8, 0, 2, 0}};

    int sol[] = new int[n+1];

    GraphAlgo.quadraticAssignment(n,a,b,sol);
    System.out.println("Optimal assignment:");
    for (int i=1; i<=n; i++)
      System.out.println("  " + i + " -- " + sol[i]);
    System.out.println("\nOptimal objective function value = " + sol[0]);
  }
}
```

Output:

```
Optimal assignment:
  1 -- 4
  2 -- 6
  3 -- 3
  4 -- 2
  5 -- 1
  6 -- 5
  7 -- 7

Optimal objective function value = 374
```

9.4 Multiple Knapsack Problem

The Multiple Knapsack problem is a generalization of the single knapsack problem and is defined as follows. We are given a set of n items and m knapsacks (bins) such that each item j has a profit p_j and a weight w_j associated with it, and each knapsack i has its own capacity c_i. The goal is to pack items into the knapsacks so as to maximize the profit of packed items without exceeding the capacity of each knapsack. This can be formulated as an optimization problem:

$$\text{maximize} \quad \sum_{i=1}^{m} \sum_{j=1}^{n} p_j x_{ij}$$

subject to

$$\sum_{j=1}^{n} w_j x_{ij} \le c_i \qquad (i = 1,2,\ldots,m)$$

$$\sum_{i=1}^{m} x_{ij} \le 1 \qquad (j = 1,2,\ldots,n)$$

$$x_{ij} = 0 \text{ or } 1 \qquad (i = 1,2,\ldots,m, \quad j = 1,2,\ldots,n)$$

where x_{ij} has the value 1 if and only if item j is assigned to knapsack i.

The following procedure [MT85] solves the multiple knapsack problem through a depth-first tree-search enumerative scheme. Each node of the decision tree produces two branches. One branch corresponds to the assignment of an item j to a knapsack i and the other branch corresponds to the exclusion of item j from knapsack i. The worst case running time of the algorithm in finding an optimal solution can be exponential. The implementation supports the option of terminating the procedure after a specified number of backtrack iterations and the best solution obtained so far will be returned as an approximate solution.

Procedure parameters:

void multipleKnapsack (*n, m, profit, weight, capacity, depth, sol*)

n: int;
entry: the number of items.

m: int;
entry: the number of knapsacks, m ≥ 1.

profit: int[*n*+1];
entry: *profit[i]* is the positive profit of item i, i=1,2,...,*n*.

weight: int[*n*+1];
entry: *weight[i]* is the positive weight of item i, i=1,2,...,*n*.

capacity: int[*m*+1];
entry: *capacity[i]* is the positive capacity of knapsack i, i=1,2,...,*m*; this input array must be sorted in nondecreasing order, that is, *capacity[i]* ≤ *capacity[i+1]* for i=1,2,...,*m*–1.

depth: int;
entry: if the optimal solution is required, then set the value to be –1; if an approximate solution is required then set the value to be the maximum number of backtrack iterations to be performed.

sol: int[n+1];
exit: if *sol[0]* is positive then an optimal solution is found; the optimal total profit is stored in *sol[0]*, and *sol[i]* is the knapsack in which item i is packed, for i=1,2,...,n. If *sol[0]* is negative, then *sol[0]* will have one of the following return values:
–1: either $n < 2$ or $m < 1$.
–2: some input value of *profit, weight,* or *capacity* is not positive.
–3: a knapsack is too small to contain any item.
–4: array *capacity* is not sorted in nondecreasing order.

```
public static void multipleKnapsack(int n, int m, int profit[], int weight[],
                                    int capacity[], int depth, int sol[])
{
  int i, i1, p, idx, idx1, idx2, j, y, tmp, netp, slackbnd, res1a, ubtmp;
  int minweight, totalweight, depthtmp, n1, m1, proftmp, ubslack;
  int q=0, upperbnd=0, indexj=0, indexi=0;
  float r;
  float pwratio[] = new float[n + 1];
  int res1[] = new int[1];
  int res2[] = new int[1];
  int res3[] = new int[1];
  int res4[] = new int[1];
  int totalprofit[] = new int[1];
  int origp[] = new int[n + 1];
  int origw[] = new int[n + 1];
  int origindex[] = new int[n + 1];
  int aux0[] = new int[m + 1];
  int aux1[] = new int[m + 1];
  int aux2[] = new int[m + 1];
  int aux3[] = new int[m + 1];
  int aux4[] = new int[n + 1];
  int aux5[] = new int[n + 1];
  int aux6[] = new int[n + 1];
  int aux7[] = new int[n + 1];
  int aux8[] = new int[n + 1];
  int aux9[] = new int[n + 2];
  int aux10[] = new int[n + 2];
  int aux11[] = new int[n + 2];
  int aux12[][] = new int[m+1][n + 1];
  int aux13[][] = new int[m+1][n+1];
  int aux14[][] = new int[m+1][n+1];
  int aux15[][] = new int[m+1][n+2];
  boolean control[] = new boolean[1];
  boolean skip=false, outer=false;
  boolean unmark[] = new boolean[n + 1];

  // check for invalid input data
  totalprofit[0] = 0;
  if (n <= 1) totalprofit[0] = - 1;
```

```
if (m <= 0) totalprofit[0] = - 1;
if (totalprofit[0] < 0) {
  sol[0] = totalprofit[0];
  return;
}
minweight = weight[1];
totalweight = weight[1];
if (profit[1] <= 0) totalprofit[0] = -2;
if (weight[1] <= 0) totalprofit[0] = -2;
for (j=2; j<=n; j++) {
  if (profit[j] <= 0) totalprofit[0] = -2;
  if (weight[j] <= 0) totalprofit[0] = -2;
  if (weight[j] < minweight) minweight = weight[j];
  totalweight += weight[j];
}

// store the original input
for (j=1; j<=n; j++) {
  origp[j] = profit[j];
  origw[j] = weight[j];
  pwratio[j] = ((float) profit[j]) / ((float) weight[j]);
  unmark[j] = true;
}

// sort the input
for (i=1; i<=n; i++) {
  r = -1.0F;
  for (j=1; j<=n; j++)
    if (unmark[j]) {
      if (pwratio[j] > r) {
        r = pwratio[j];
        q = j;
      }
    }
  unmark[q] = false;
  profit[i] = origp[q];
  weight[i] = origw[q];
  origindex[i] = q;
}

if (capacity[1] <= 0) totalprofit[0] = -2;
if (m == 1) {
  // solve the special case of one knapsack problem
  if (minweight > capacity[1]) totalprofit[0] = -3;
  if (totalweight <= capacity[1]) {
    // the knapsacks contain all the items
    q = 0;
    for (j=1; j<=n; j++) {
      profit[j] = origp[j];
      weight[j] = origw[j];
```

```
        sol[j] = 1;
        q += origw[j];
      }
      sol[0] = q;
      return;
    }
    if (totalprofit[0] < 0) {
      sol[0] = totalprofit[0];
       for (j=1; j<=n; j++) {
        profit[j] = origp[j];
        weight[j] = origw[j];
      }
     return;
    }
    res4[0] = capacity[1];
    for (j=1; j<=n; j++) {
      aux10[j] = profit[j];
      aux11[j] = weight[j];
    }
    // compute the solution with one knapsack
    mkpsSingleKnapsack(n,n,res4,0,totalprofit,aux8,aux10,aux11);
    depth = 0;
    sol[0] = totalprofit[0];
    for (j=1; j<=n; j++) {
      profit[j] = origp[j];
      weight[j] = origw[j];
      sol[origindex[j]] = aux8[j];
    }
    return;
  }
  for (i=2; i<=m; i++) {
    if (capacity[i] <= 0) totalprofit[0] = -2;
    if (capacity[i] < capacity[i-1]) {
      sol[0] = -4;
      for (j=1; j<=n; j++) {
        profit[j] = origp[j];
        weight[j] = origw[j];
      }
      return;
    }
  }
  if (minweight > capacity[1]) totalprofit[0] = -3;
  if (totalweight <= capacity[m]) totalprofit[0] = -4;
  if (totalprofit[0] < 0) {
    sol[0] = totalprofit[0];
    for (j=1; j<=n; j++) {
      profit[j] = origp[j];
      weight[j] = origw[j];
    }
    return;
```

```
  }

  // initialize
  depthtmp = depth;
  depth = 0;
  netp = 0;
  n1 = n + 1;
  aux9[n1] = 1;
  m1 = m - 1;
  for (j=1; j<=n; j++) {
    aux9[j] = 1;
    for (i=1; i<=m; i++) {
      aux13[i][j] = 0;
      aux12[i][j] = 0;
    }
  }
  for (i=1; i<=m1; i++) {
    aux2[i] = capacity[i];
    aux0[i] = -1;
  }
  aux2[m] = capacity[m];
  totalprofit[0] = 0;
  slackbnd = 0;
  idx = 1;
  // compute an upper bound of the current solution
  mkpsCurrentUpperBound(n,m,profit,weight,capacity,1,netp,res1,res3,
        aux5,aux7,aux8,aux9,aux10,aux11);
  for (j=1; j<=n; j++)
    aux6[j] = aux5[j];
  res1a = res1[0];
  ubtmp = res3[0];
  control[0] = false;

  while (true) {
    // using heuristic approximation
    outer = false;
    netp = totalprofit[0] - slackbnd;
    // get a feasible solution
    mkpsFeasibleSolution(n,m,profit,weight,idx,netp,res2,aux1,aux2,
                 aux3,aux7,aux8,aux9,aux10,aux11,aux12,aux14,aux15);
    if (res2[0] + slackbnd > totalprofit[0]) {
      totalprofit[0] = res2[0] + slackbnd;
      for (j=1; j<=n; j++) {
        sol[j] = 0;
        for (y=1; y<=idx; y++)
          if (aux13[y][j] != 0) {
            sol[j] = y;
            break;
          }
      }
```

```
  idx1 = aux1[i];
  if (idx1 != 0) {
    for (j=1; j<=idx1; j++) {
      q = aux15[idx][j];
      if (aux14[idx][j] == 1) sol[q] = idx;
    }
  }
  i1 = idx + 1;
  for (p=i1; p<=m; p++) {
    idx1 = aux1[p];
    if (idx1 != 0)
      for (j=1; j<=idx1; j++) {
        q = aux15[p][j];
        if (aux14[p][j] == 1) sol[q] = p;
      }
  }
  if (res3[0] == res2[0]) {
    outer = true;
  }
}
if (!outer) {
  skip = false;
  while (true) {
    if (aux3[idx] != 0) {
      // update
      ubslack = res3[0] + slackbnd;
      tmp = aux1[idx];
      proftmp = 0;
      for (y=1; y<=tmp; y++) {
        if (aux14[idx][y] != 0) {
          j = aux15[idx][y];
          aux13[idx][j] = 1;
          aux2[idx] -= weight[j];
          slackbnd += profit[j];
          aux9[j] = 0;
          aux12[idx][j] = aux0[idx];
          aux4[j] = ubslack;
          if (!control[0]) {
            upperbnd = ubslack;
            indexj = j;
            indexi = idx;
          }
          aux0[idx] = j;
          proftmp += profit[j];
          if (proftmp == aux3[idx]) {
            skip = true;
            break;
          }
          // upper bound computation
          mkpsCalculateBound(idx,idx,res3,control,slackbnd,upperbnd,
```

```
                              indexj,indexi,n,res1,res1a,ubtmp,aux0,
                              aux2,aux5,aux6,aux9,aux12);
          if (!control[0]) {
            netp = totalprofit[0] - slackbnd;
            // compute an upper bound of the current solution
            mkpsCurrentUpperBound(n,m,profit,weight,aux2,idx,netp,
                        res1,res3,aux5,aux7,aux8,aux9,aux10,aux11);
            indexj = n1;
          }
          ubslack = res3[0] + slackbnd;
          if (ubslack <= totalprofit[0]) {
            outer = true;
            break;
          }
        }
      }
      if (skip || outer) break;
    }
    if (idx == m - 1) {
      outer = true;
      break;
    }
    idx2 = idx + 1;
    // upper bound computation
    mkpsCalculateBound(idx2,idx,res3,control,slackbnd,upperbnd,indexj,
            indexi,n,res1,res1a,ubtmp,aux0,aux2,aux5,aux6,aux9,aux12);
    if (!control[0]) {
      netp = totalprofit[0] - slackbnd;
      // compute an upper bound of the current solution
      mkpsCurrentUpperBound(n,m,profit,weight,aux2,idx2,netp,res1,
                            res3,aux5,aux7,aux8,aux9,aux10,aux11);
      indexj = n1;
    }
    if (res3[0] + slackbnd <= totalprofit[0]) {
      outer = true;
      break;
    }
    idx++;
  }
}
while (true) {
  // backtrack
  if (idx <= 0) {
    depth--;
    sol[0] = totalprofit[0];
    for (j=1; j<=n; j++)
      aux8[j] = sol[j];
    for (j=1; j<=n; j++) {
      profit[j] = origp[j];
      weight[j] = origw[j];
```

```
          sol[origindex[j]] = aux8[j];
        }
        return;
      }
      if (depth == depthtmp) {
        sol[0] = totalprofit[0];
        for (j=1; j<=n; j++)
          aux8[j] = sol[j];
        for (j=1; j<=n; j++) {
          profit[j] = origp[j];
          weight[j] = origw[j];
          sol[origindex[j]] = aux8[j];
        }
        return;
      }
      depth++;
      if (aux0[idx] == -1) {
        for (j=1; j<=n; j++)
          aux12[idx][j] = 0;
        idx--;
        continue;
      }
      j = aux0[idx];
      aux13[idx][j] = 0;
      aux9[j] = 1;
      slackbnd -= profit[j];
      aux2[idx] += weight[j];
      for (y=1; y<=n; y++)
        if (aux12[idx][y] == j) aux12[idx][y] = 0;
      aux0[idx] = aux12[idx][j];
      if (aux4[j] > totalprofit[0]) break;
    }
    res3[0] = aux4[j] - slackbnd;
    control[0] = true;
  }
}


static private void mkpsCurrentUpperBound(int n, int m, int profit[],
       int weight[], int aux2[], int i, int netp, int res1[],
       int res3[], int aux5[], int aux7[], int aux8[], int aux9[],
       int aux10[], int aux11[])
{
  /* this method is used internally by multipleKnapsack */

  // Compute an upper bound of the current solution

  int j, q, wk1, wk2;
  int ref1[] = new int[1];
```

```
  wk1 = 0;
  ref1[0] = 0;
  for (j=i; j<=m; j++)
    ref1[0] += aux2[j];
  wk2 = 0;
  for (j=1; j<=n; j++) {
    aux5[j] = 0;
    if (aux9[j] != 0) {
      wk1++;
      aux7[wk1] = j;
      aux10[wk1] = profit[j];
      aux11[wk1] = weight[j];
      wk2 += weight[j];
    }
  }
  if (wk2 <= ref1[0]) {
    res1[0] = ref1[0] - wk2;
    res3[0] = 0;
    if (wk1 == 0) return;
    for (j=1; j<=wk1; j++) {
      res3[0] += aux10[j];
      aux8[j] = 1;
    }
  }
  else {
    // compute the solution with one knapsack
    mkpsSingleKnapsack(n,wk1,ref1,netp,res3,aux8,aux10,aux11);
    res1[0] = ref1[0];
  }
  for (j=1; j<=wk1; j++) {
    q = aux7[j];
    aux5[q] = aux8[j];
  }
}


static private void mkpsFeasibleSolution(int n, int m, int profit[],
       int weight[], int i, int netp, int res2[], int aux1[], int aux2[],
       int aux3[], int aux7[], int aux8[], int aux9[], int aux10[],
       int aux11[], int aux12[][], int aux14[][], int aux15[][])
{
  /* this method is used internally by multipleKnapsack */

  // Get a feasible solution

  int p,j,q,netpa,accu1,accu2,accu3,accu4,accu5;
  int ref1[] = new int[1];
  int pb[] = new int[1];

  accu5 = 0;
```

```
for (j=1; j<=n; j++)
  if (aux9[j] != 0) {
    accu5++;
    aux7[accu5] = j;
  }
for (j=i; j<=m; j++) {
  aux1[j] = 0;
  aux3[j] = 0;
}
res2[0] = 0;
netpa = netp;
if (accu5 == 0) return;
accu3 = 0;
accu4 = 0;
for (j=1; j<=accu5; j++) {
  q = aux7[j];
  if (aux12[i][q] == 0) {
    if (weight[q] <= aux2[i]) {
      accu3++;
      accu4 += weight[q];
      aux15[i][accu3] = q;
      aux10[accu3] = profit[q];
      aux11[accu3] = weight[q];
    }
  }
}
p = i;
while (true) {
  aux1[p] = accu3;
  if (accu4 <= aux2[p]) {
    pb[0] = 0;
    if (accu3 != 0) {
      for (j=1; j<=accu3; j++) {
        pb[0] += aux10[j];
        aux14[p][j] = 1;
      }
    }
  }
  else {
    ref1[0] = aux2[p];
    netp = 0;
    if (p == m) netp = netpa;
    // compute the solution with one knapsack
    mkpsSingleKnapsack(n,accu3,ref1,netp,pb,aux8,aux10,aux11);
    for (j=1; j<=accu3; j++)
      aux14[p][j] = aux8[j];
  }
  res2[0] += pb[0];
  netpa -= pb[0];
  aux3[p] = pb[0];
```

```
    aux15[p][accu3+1] = n + 1;
    if (p == m) return;
    accu1 = 1;
    accu2 = 0;
    for (j=1; j<=accu5; j++) {
      if (aux7[j] >= aux15[p][accu1]) {
        accu1++;
        if (aux14[p][accu1-1] == 1) continue;
      }
      accu2++;
      aux7[accu2] = aux7[j];
    }
    accu5 = accu2;
    if (accu5 == 0) return;
    accu3 = 0;
    accu4 = 0;
    p++;
    for (j=1; j<=accu5; j++) {
      q = aux7[j];
      if(weight[q] <= aux2[p]) {
        accu3++;
        accu4 += weight[q];
        aux15[p][accu3] = q;
        aux10[accu3] = profit[q];
        aux11[accu3] = weight[q];
      }
    }
  }
}


static private void mkpsCalculateBound(int i, int p, int res3[],
       boolean control[], int slackbnd, int upperbnd, int indexj,
       int indexi, int n, int res1[], int res1a, int ubtmp,
       int aux0[], int aux2[], int aux5[], int aux6[], int aux9[],
       int aux12[][])
{
  /* this method is used internally by multipleKnapsack */

  // Upper bound computation

  int j,id1,id2,id3,id4;

  control[0] = false;
  if (aux9[indexj] == 0) {
    id1 = i - 1;
    if (id1 >= indexi) {
      id2 = 0;
      for (j=indexi; j<=id1; j++)
        id2 += aux2[j];
```

```
      if (id2 > res1[0]) return;
    }
    id3 = p;
    id4 = aux0[id3];
    while (true) {
      if (id4 == -1) {
        id3--;
        id4 = aux0[id3];
      }
      else {
        if (aux5[id4] == 0) return;
        if (id4 == indexj) break;
        id4 = aux12[id3][id4];
      }
    }
    res3[0] = upperbnd - slackbnd;
    control[0] = true;
    return;
  }
  id1 = i - 1;
  if (id1 >= 1) {
    id2 = 0;
    for (j=1; j<=id1; j++)
      id2 +- aux2[j];
    if (id2 > res1a) return;
  }
  for (j=1; j<=n; j++)
    if (aux9[j] != 1)
      if (aux6[j] == 0) return;
  res3[0] = ubtmp - slackbnd;
  control[0] = true;
}


static private void mkpsSingleKnapsack(int n, int ns, int ref1[],
       int netp,int ref2[], int aux8[], int aux10[], int aux11[])
{
  /* this method is used internally by multipleKnapsack */

  // Compute the solution with one knapsack

  int p,p1,in,j,j1,q,diff,index2,index3,index4,index5,index6;
  int thres,thres1,val2,prev,n1,r;
  int val1=0,index1=0,index7=0,t=0;
  float tmp1,tmp2,tmp3;
  int work1[] = new int[n + 1];
  int work2[] = new int[n + 1];
  int work3[] = new int[n + 1];
  int work4[] = new int[n + 1];
  int work5[] = new int[n + 1];
```

```
boolean skip=false, jump=false, middle=false, over=false, outer=false;

ref2[0] = netp;
index3 = 0;
index2 = ref1[0];
for (j=1; j<=ns; j++) {
  index1 = j;
  if (aux11[j] > index2) break;
  index3 += aux10[j];
  index2 -= aux11[j];
}
index1--;
if (index2 != 0) {
  aux10[ns+1] = 0;
  aux11[ns+1] = ref1[0] + 1;
  thres = index3 + index2 * aux10[index1+2] / aux11[index1+2];
  tmp1 = index3 + aux10[index1+1];
  tmp2 = (aux11[index1+1] - index2) * aux10[index1];
  tmp3 = aux11[index1];
  thres1 = (int) (tmp1 - tmp2 / tmp3);
  if (thres1 > thres) thres = thres1;
  if (thres <= ref2[0]) return;
  val2 = ref1[0] + 1;
  work2[ns] = val2;
  for (j=2; j<=ns; j++) {
    index7 = ns + 2 - j;
    if (aux11[index7] < val2) val2 = aux11[index7];
    work2[index7-1] = val2;
  }
  for (j=1; j<=ns; j++)
    work1[j] = 0;
  index5 = 0;
  prev = ns;
  p = 1;
  skip = true;
}
else {
  if (ref2[0] >= index3) return;
  ref2[0] = index3;
  for (j=1; j<=index1; j++)
    aux8[j] = 1;
  index6 = index1 + 1;
  for (j=index6; j<=ns; j++)
    aux8[j] = 0;
  ref1[0] = 0;
  return;
}
middle = false;
while (true) {
  if (!skip) {
```

```
    if (aux11[p] > ref1[0]) {
      p1 = p + 1;
      if (ref2[0] >= ref1[0] * aux10[p1] / aux11[p1] + index5) {
        middle = true;
      }
      if (!middle) {
        p = p1;
        continue;
      }
    }
    if (!middle) {
      index3 = work3[p];
      index2 = ref1[0] - work4[p];
      in = work5[p];
      index1 = ns;
      if (in <= ns) {
        for (j=in; j<=ns; j++) {
          index1 = j;
          if (aux11[j] > index2) {
            index1--;
            if (index2 == 0) break;
            if (ref2[0] >= index5 + index3 + index2 * aux10[j] / aux11[j]) {
              middle = true;
              break;
            }
            skip = true;
            break;
          }
          index3 += aux10[j];
          index2 -= aux11[j];
        }
      }
    }
    if (!middle) {
      if (!skip) {
        if (ref2[0] >= index3 + index5) {
          middle = true;
        }
        if (!middle) {
          ref2[0] = index3 + index5;
          val1 = index2;
          index6 = p - 1;
          for (j=1; j<=index6; j++)
            aux8[j] = work1[j];
          for (j=p; j<=index1; j++)
            aux8[j] = 1;
          if (index1 != ns) {
            index6 = index1 + 1;
            for (j=index6; j<=ns; j++)
              aux8[j] = 0;
```

```
            }
            if (ref2[0] != thres) {
              middle = true;
            }
            if (!middle) {
              ref1[0] = val1;
              return;
            }
          }
        }
      }
    }
    if (!middle) {
      skip = false;
      work4[p] = ref1[0] - index2;
      work3[p] = index3;
      work5[p] = index1 + 1;
      work1[p] = 1;
      index6 = index1 - 1;
      if (index6 >= p)
        for (j=p; j<=index6; j++) {
          work4[j+1] = work4[j] - aux11[j];
          work3[j+1] = work3[j] - aux10[j];
          work5[j+1] = index1 + 1;
          work1[j+1] = 1;
        }
      j1 = index1 + 1;
      for (j=j1; j<=prev; j++) {
        work4[j] = 0;
        work3[j] = 0;
        work5[j] = j;
      }
      prev = index1;
      ref1[0] = index2;
      index5 += index3;
      if ((index1 - (ns - 2)) > 0)
        p = ns;
      else if ((index1 - (ns - 2)) == 0) {
        if (ref1[0] >= aux11[ns]) {
          ref1[0] -= aux11[ns];
          index5 += aux10[ns];
          work1[ns] = 1;
        }
        p = ns - 1;
      }
      else {
        p = index1 + 2;
        if (ref1[0] >= work2[p-1]) continue;
      }
      if (ref2[0] < index5) {
```

```
        ref2[0] = index5;
        for (j=1; j<=ns; j++)
          aux8[j] = work1[j];
        val1 = ref1[0];
        if (ref2[0] == thres) return;
      }
      if (work1[ns] != 0) {
        work1[ns] = 0;
        ref1[0] += aux11[ns];
        index5 -= aux10[ns];
      }
    }
    outer = false;
    while (true) {
      middle = false;
      index6 = p - 1;
      jump = false;
      if (index6 != 0) {
        for (j=1; j<=index6; j++) {
          index7 = p - j;
          if (work1[index7] == 1) {
            jump = true;
            break;
          }
        }
      }
      if (!jump) {
        ref1[0] = val1;
        return;
      }
      r = ref1[0];
      ref1[0] += aux11[index7];
      index5 -= aux10[index7];
      work1[index7] = 0;
      if (r >= work2[index7]) {
        p = index7 + 1;
        outer = true;
        break;
      }
      index6 = index7 + 1;
      p = index7;
      over = false;
      while (true) {
        if (ref2[0] >= index5 + ref1[0] * aux10[index6] / aux11[index6]) {
          over = true;
          break;
        }
        diff = aux11[index6] - aux11[index7];
        if (diff < 0) {
          t = r - diff;
```

```
          if (t < work2[index6]) {
            index6++;
            continue;
          }
          break;
        }
        if (diff == 0) {
          index6++;
          continue;
        }
        if (diff > r) {
          index6++;
          continue;
        }
        if (ref2[0] >= index5 + aux10[index6]) {
          index6++;
          continue;
        }
        ref2[0] = index5 + aux10[index6];
        for (j=1; j<=index7; j++)
          aux8[j] = work1[j];
        q = index7 + 1;
        for (j=q; j<=ns; j++)
          aux8[j] = 0;
        aux8[index6] = 1;
        val1 = ref1[0] - aux11[index6];
        if (ref2[0] == thres) {
          ref1[0] = val1;
          return;
        }
        r -= diff;
        index7 = index6;
        index6++;
      }
      if (!over) {
        n = index6 + 1;
        if (ref2[0] < index5 + aux10[index6] + t * aux10[n] / aux11[n])
          break;
      }
      else
        over = false;
    }
    if (!outer) {
      ref1[0] -= aux11[index6];
      index5 += aux10[index6];
      work1[index6] = 1;
      p = index6 + 1;
      work4[index6] = aux11[index6];
      work3[index6] = aux10[index6];
      work5[index6] = p;
```

```
      n1 = index6 + 1;
      for (j=n1; j<=prev; j++) {
        work4[j] = 0;
        work3[j] = 0;
        work5[j] = j;
      }
      prev = index6;
    }
    else
      outer = false;
  }
}
```

Example:

Solve the multiple knapsack problem of 2 knapsacks and 10 items with the following weights and profits:

Profit: 46, 50, 45, 58, 74, 52, 36, 30, 79, 61

Weight: 81, 83, 43, 34, 68, 58, 60, 72, 42, 28

The capacity of the two knapsacks are 125 and 146, respectively.

```
package GraphAlgorithms;

public class Test_knapsack extends Object {

  public static void main(String args[]) {

    int n = 10;
    int m = 2;
    int depth = -1;
    int profit[] = {0, 46, 50, 45, 58, 74, 52, 36, 30, 79, 61};
    int weight[] = {0, 81, 83, 43, 34, 68, 58, 60, 72, 42, 28};
    int capacity[] = {0, 125, 146};
    int sol[] = new int[n + 1];

    GraphAlgo.multipleKnapsack(n, m, profit, weight, capacity, depth, sol);
    if (sol[0] > 0) {
      System.out.println("Optimal solution found:");
      for (int i=1; i<=n; i++)
        System.out.print("  " + sol[i]);
      System.out.println("\n\nTotal profit = " + sol[0]);
    }
    else
      System.out.println("Error returned = " + sol[0]);
  }
}
```

Output:

```
Optimal solution found:
  0  0  2  1  0  1  2  0  2  1

Total profit = 331
```

9.5 Set Covering Problem

Consider a set Q = {1,2,...,m}, and a set { S_1, S_2, ..., S_n }, where $S_j \subseteq Q$, $j \in P = \{1,2,...,n\}$. A subet $R \subseteq P$ is a *cover* of Q if

$$\bigcup_{j \in R} S_j = Q$$

Let $c_j > 0$ be the cost associated with every $j \in P$. The total cost of the cover R is

$$\sum_{j \in R} c_j$$

The *minimum cost set covering problem* is to find a cover of minimum cost, and can be formulated as a zero-one integer programming problem:

$$\text{minimize} \quad \sum_{j=1}^{n} c_j x_j$$

$$\text{subject to} \quad \sum_{j=1}^{n} a_{ij} x_j \geq 1, \qquad (i = 1,2,\ldots,m)$$

$$x_j = 0 \text{ or } 1, \qquad (j = 1,2,\ldots,n)$$

where x_j = 1 if j is in the cover, x_j = 0 otherwise; and a_{ij} = 1 if $i \in S_j$, a_{ij} = 0 otherwise.

Consider the special case when c_j = 1, for j=1,2,...,n. The size of R is the number of elements in R. The *set covering problem* is to find a cover R of minimum size.

The solution of the *minimum cost set covering problem* and the *set covering problem* will be obtained by solving its associated zero-one integer programming problem (see Chapter 11 for the zero-one integer programming code).

Procedure parameters:

void setCovering (*n, m, a, c, sol*)

n: int;
entry: number of subsets described above.

m: int;
entry: number of elements of the set Q described above.

a: double[*m*+1][*n*+1];
entry: $a_{ij} = 1$ if $i \in S_j$, $a_{ij} = 0$ otherwise, i=1,2,...,*m*, j=1,2,...,*n*.
The other elements of the matrix *a* are not required as input.
exit: if an optimal solution is found then *a[0][0]* = 0;
if there is no feasible solution then *a[0][0]* > 0.

c: int[*n*+1];
entry: *c[j]* > 0 is the cost associated with every $j \in P$ described above, j=1,2,...,*n*.

sol: int[*n*+1];
exit: in the optimal solution, *sol[j]* = 1 if j is in the cover, *sol[j]* = 0 otherwise, *j* = 1, 2, ..., *n*.
sol[0] is the optimal value of the objective function.

```
public static void setCovering(int n, int m, int a[][],
                               int c[], int sol[])
{
  int i,j;
  int b[] = new int[m + 1];

  for (i=0; i<=m; i++)
    for (j=0; j<=n; j++)
      a[i][j] = (a[i][j] == 1) ? -1 : 0;
  for (i=0; i<=m; i++)
    b[i] = -1;
  zeroOneIntegerProgramming(true, n, m, a, b, c, sol);
}
```

Example:

Find a minimum cost set cover for the following problem:
4 elements (m = 4), and 5 subsets (n = 5),
$S_1 = \{1, 3\}$
$S_2 = \{2\}$
$S_3 = \{1, 4\}$
$S_4 = \{1, 2\}$
$S_5 = \{3, 4\}$
associated cost vector c = (7, 2, 5, 8, 6).

```
package Optimization;

public class Test_setCovering extends Object {

  public static void main(String args[]) {

    int n = 5;
    int m = 4;
    int a[][] = new int[m+1][n+1];
    int c[] = {0, 7, 2, 5, 8, 6};
```

```
    int sol[] = new int[n + 1];

    a[1][1] = a[1][3] = a[1][4] = 1;
    a[2][2] = a[2][4] = 1;
    a[3][1] = a[3][5] = 1;
    a[4][3] = a[4][5] = 1;
    Optimize.setCovering(n, m, a, c, sol);
    if (a[0][0] > 0)
      System.out.println("No feasible solution.");
    else {
      System.out.print("Optimal solution found.\n\n Solution vector: ");
      for (int i=1; i<=n; i++)
        System.out.print("   " + sol[i]);
      System.out.println("\n\nOptimal objective function value = " + sol[0]);
    }
  }
}
```

Output:

```
Optimal solution found.

 Solution vector:   0  1  1  0  1

Optimal objective function value = 13
```

9.6 Set Partitioning Problem

Consider a set Q = {1,2,...,m}, and a set C = { S_1, S_2, ..., S_n }, where $S_j \subseteq Q$, $j \in P = \{1,2,\ldots,n\}$. A subet $R \subseteq P$ is a *partition* of Q if

$$\bigcup_{j\in R} S_j = Q$$

and, in addition, $u, v \in R$, $u \neq v \Rightarrow S_u \cap S_v = \varnothing$.
Let $c_j > 0$ be the cost associated with every $j \in P$. The total cost of the partition R is

$$\sum_{j\in R} c_j$$

The *minimum cost set partitioning problem* is to find a partition of minimum cost, and can be written as the zero-one integer programming problem:

$$\text{minimize} \quad \sum_{j=1}^{n} c_j x_j$$

$$\text{subject to} \quad \sum_{j=1}^{n} a_{ij} x_j = 1, \qquad (i = 1,2,\ldots,m)$$

$$x_j = 0 \text{ or } 1, \qquad (j = 1,2,\ldots,n)$$

where x_j = 1 if j is in the partition, x_j = 0 otherwise; and a_{ij} = 1 if i ∈ S_j, a_{ij} = 0 otherwise.

Note that the *minimum cost set covering problem* is more general than the *minimum cost set partition problem*. Indeed any *minimum cost set partitioning problem* having a feasible solution can be transformed into a *minimum cost set covering problem* by changing the cost vector. More precisely, let

$$w_j = \sum_{i=1}^{m} a_{ij} \quad \text{and choose any } K > \sum_{j=1}^{n} c_j$$

if the *minimum cost set partitioning problem* defined by $A = \{a_{ij}\}$ and costs c = $\{c_j\}$ has a partition, then the *minimum cost set covering problem* defined by $A = \{a_{ij}\}$ and costs $\hat{c} = c_j + Kw_j$, j=1,2,...,n, has the same set of optimal solutions as the *minimum cost set partitioning problem.*

The solution of the *minimum cost set partitioning problem* will be obtained by solving the associated zero-one integer programming problem (see Chapter 11 for the zero-one integer programming code) of the transformed minimum cost set covering problem.

Procedure parameters:

void setPartitioning (*n, m, a, c, sol*)

n: int;
entry: number of subsets described above.

m: int;
entry: number of elements of the set Q described above.

a: double[*m*+1][*n*+1];
entry: a_{ij} = 1 if i ∈ S_j, a_{ij} = 0 otherwise, *i*=1,2,...,*m*, *j*=1,2,...,*n*.
The other elements of the matrix *a* are not required as input.
exit: if an optimal solution is found then *a[0][0]* = 0;
if there is no feasible solution then *a[0][0]* > 0.

c: int[*n*+1];
entry: *c[j]* > 0 is the cost associated with every j ∈ P described above, *j*=1,2,...,*n*.

sol: int[*n*+1];
exit: in the optimal solution, *sol[j]* = 1 if j is in the partition, *sol[j]* = 0 otherwise, *j* = 1, 2, ..., *n*.
sol[0] is the optimal value of the objective function.

```
public static void setPartitioning(int n, int m, int a[][], int c[], int sol[])
{
  int i,j,sum,mult;
  int b[] = new int[m + 1];
  int newc[] = new int[n+1];

  for (i=0; i<=m; i++)
    for (j=0; j<=n; j++)
```

```
      a[i][j] = (a[i][j] == 1) ? -1 : 0;
  for (i=0; i<=m; i++)
    b[i] = -1;

  // transform the cost vector, original cost vector not changed
  sum = 0;
  for (j=1; j<=n; j++) {
    sum += c[j];
    newc[j] = c[j];
  }
  mult = sum + 1;
  for (j=1; j<=n; j++) {
    sum = 0;
    for (i=1; i<=m; i++)
      sum += a[i][j];
    newc[j] += mult*(-sum);
  }
  // use the transformed cost vector to solve the set partitioning problem
  Optimize.zeroOneIntegerProgramming(true, n, m, a, b, newc, sol);
  if (a[0][0] == 0) {
  //  obtain the optimal objective function value by the solution vector
    sum = 0;
    for (i=1; i<=n; i++) {
      sum += (sol[i] == 0) ? 0 : c[i];
    }
    sol[0] = sum;
  }
}
```

Example:

Find a minimum cost set partition for the following problem:

4 elements (m = 4), and 6 subsets (n = 6),
$S_1 = \{1, 4\}$
$S_2 = \{2, 4\}$
$S_3 = \{1, 2, 3\}$
$S_4 = \{3, 4\}$
$S_5 = \{2, 3\}$
$S_6 = \{1, 2\}$
associated cost vector c = (4, 2, 1, 6, 5, 2).

```
package Optimization;

public class Test_setPartitioning extends Object {

  public static void main(String args[]) {

    int n = 6;
    int m = 4;
    int a[][] = new int[m+1][n+1];
    int c[] = {0, 4, 2, 1, 6, 5, 2};
    int sol[] = new int[n + 1];

    a[1][1] = a[1][3] = a[1][6] = 1;
    a[2][2] = a[2][3] = a[2][5] = a[2][6] = 1;
    a[3][3] = a[3][4] = a[3][5] = 1;
    a[4][1] = a[4][2] = a[4][4] = 1;
    Optimize.setPartitioning(n, m, a, c, sol);
    if (a[0][0] > 0)
      System.out.println("No feasible solution.");
    else {
      System.out.print("Optimal solution found\n\n Solution vector: ");
      for (int i=1; i<=n; i++)
        System.out.print("  " + sol[i]);
      System.out.println("\n\nOptimal objective function value = " + sol[0]);
    }
  }
}
```

Output:

```
Optimal solution found

 Solution vector:   0  0  0  1  0  1

Optimal objective function value = 8
```

10. Linear Programming

10.1 Revised Simplex Method

The following is a *linear programming problem* in standard form:

$$\text{maximize} \quad \sum_{j=1}^{n} c_j x_j$$

$$\text{subject to} \quad \sum_{j=1}^{n} a_{ij} x_j \le b_i \qquad (i = 1,2,\ldots,m)$$

$$x_j \ge 0 \qquad (j = 1,2,\ldots,n)$$

After introducting the slack variables x_{n+1}, x_{n+2}, ..., x_{n+m}, the problem can be written as:

$$\text{maximize} \quad \sum_{j=1}^{n} c_j x_j$$

$$\text{subject to} \quad \sum_{j=1}^{n} a_{ij} x_j = b_i \qquad (i = 1,2,\ldots,m)$$

$$x_j \ge 0 \qquad (j = 1,2,\ldots,n)$$

or, in matrix notation:

maximize cx
subject to Ax = b
x ≥ 0

The matrix A has m rows and n+m columns with the last m columns forming an identity matrix. The vector x is of length n+m, and the column b is of length m. The linear programming problem can be solved by the well-known *revised simplex method* [DOH54]. A basic feasible solution x^* partitions x into x_B (m basic variables) and x_N (n nonbasic variables). This corresponds to the partition of matrix A into A_B and A_N, and c into c_B and c_N. Each iteration of the revised simplex method [C83, SDK83] can be described as follows:

Step 1. Solve the system $yA_B = c_B$.

Step 2. Choose any column α of A_N such that $y\alpha$ is less than the corresponding component of c_N. If such column does not exist, then the current solution is optimal.

Step 3. Solve the system $A_B\beta = \alpha$.

Step 4. Find the largest d such that $x_B^* - d\beta \ge 0$. If no such d is found then the problem is unbounded, otherwise at least one component of $x_B^* - d\beta$ will be equal to zero and the corresponding variable leaves the basis.

Step 5. Set the entering variable to be d. Replace the values of the basic variables x_B^* by $x_B^* - d\beta$. Replace the leaving column of A_B by the entering column, and replace the leaving variable by the entering variable.

Procedure parameters:

void revisedSimplex (*maximize, n, m, a, epsilon, basicvar*)

maximize: boolean;
entry: *maximize* = true, if the objective function is to be maximized;
maximize = false, if the objective function is to be minimized.

n: int;
entry: number of variables (this includes the slack variables).

m: int;
entry: number of constraints.

a: double[m+1][n+1];
entry: the coefficients of the constraints are given by *a[i][j]*, i=1,2,...,m, j=1,2,...,n.
The coefficients of the objective function are given by *a[0][j]*, j=1,2,...,n.
The right hand side of the constraints are given by *a[i][0]*, i=1,2,...,m.
The other elements of the matrix *a* are not required as input.
exit: *a[0][0]* is the optimal value of the objective function.
a[i][0] is the optimal value of the basic variable *basicvar[i]*, for i = 1, 2, ..., m.

epsilon: double;
entry: for any real number r, if |r| < epsilon then r is equal to 0.

basicvar: int[n+m+3];
exit: if there is no feasible solution then *basicvar[m+1]* > 0 otherwise *basicvar[m+1]* = 0;
if the problem has no finite solution then *basicvar[m+2]* > 0, otherwise *basicvar[m+2]* = 0;
basicvar[i] is the basic variable in the optimal solution, for i = 1, 2, ..., m.

```
public static void revisedSimplex(boolean maximize, int n, int m,
                    double a[][], double epsilon, int basicvar[])
{
  int i,j,k,m2,p,idx=0;
  double objcoeff[] = new double[n + 1];
  double varsum[] = new double[n + 1];
  double optbasicval[] = new double[m + 3];
  double aux[] = new double [m + 3];
  double work[][] = new double[m + 3][m + 3];
  double part,sum;
  boolean infeasible,unbound,abort,out,iterate;

  if (maximize)
    for (j=1; j<=n; j++)
      a[0][j] = -a[0][j];
  infeasible = false;
  unbound = false;
  m2 = m + 2;
  p = m + 2;
  out = true;
  k = m + 1;
  for (j=1; j<=n; j++) {
    objcoeff[j] = a[0][j];
    sum = 0.0;
    for (i=1; i<=m; i++)
      sum -= a[i][j];
    varsum[j] = sum;
  }
  sum = 0.0;
  for (i=1; i<=m; i++) {
    basicvar[i] = n + i;
    optbasicval[i] = a[i][0];
    sum -= a[i][0];
  }
  optbasicval[k] = 0.0;
  optbasicval[m2] = sum;
  for (i=1; i<=m2; i++) {
    for (j=1; j<=m2; j++)
      work[i][j] = 0.0;
    work[i][i] = 1.0;
  }
  iterate = true;
  do {
    // phase 1
    if ((optbasicval[m2] >= -epsilon) && out) {
      out = false;
      p = m + 1;
    }
    part = 0.0;
    // phase 2
```

```
for (j=1; j<=n; j++) {
  sum = work[p][m+1] * objcoeff[j] + work[p][m+2] * varsum[j];
  for (i=1; i<=m; i++)
    sum += work[p][i] * a[i][j];
  if (part > sum) {
    part = sum;
    k = j;
  }
}
if (part > -epsilon) {
  iterate = false;
  if (out)
    infeasible = true;
  else
    a[0][0] = -optbasicval[p];
}
else {
  for (i=1; i<=p; i++) {
    sum = work[i][m+1] * objcoeff[k] + work[i][m+2] * varsum[k];
    for (j=1; j<=m; j++)
      sum += work[i][j] * a[j][k];
    aux[i] = sum;
  }
  abort = true;
  for (i=1; i<=m; i++)
    if (aux[i] >= epsilon) {
      sum = optbasicval[i] / aux[i];
      if (abort || (sum < part)) {
        part = sum;
        idx = i;
      }
      abort = false;
    }
  if (abort) {
    unbound  = true;
    iterate = false;
  }
  else {
    basicvar[idx] = k;
    sum = 1.0 / aux[idx];
    for (j=1; j<=m; j++)
      work[idx][j] *= sum;
    i = ((idx == 1) ? 2 : 1);
    do {
      sum = aux[i];
      optbasicval[i] -= part * sum;
      for (j=1; j<=m; j++)
        work[i][j] -= work[idx][j] * sum;
      i += ((i == idx-1) ? 2 : 1);
    } while (i <= p);
```

```
          optbasicval[idx] = part;
        }
      }
    } while (iterate);
    // return results
    basicvar[m+1] = (infeasible ? 1 : 0);
    basicvar[m+2] = (unbound ? 1 : 0);
    for (i=1; i<=m; i++)
      a[i][0] = optbasicval[i];
    if (maximize) {
      for (j=1; j<=n; j++)
        a[0][j] = -a[0][j];
      a[0][0] = -a[0][0];
    }
}
```

Example:

Solve the following linear programming problem [C83] with 7 variables and 4 constraints by the revised simplex method:

maximize $5x_1 + 5x_2 + 3x_3$

subject to

$$
\begin{aligned}
x_1 + 3x_2 + x_3 + x_4 &= 3 \\
-x_1 + 3x_3 + x_5 &= 2 \\
2x_1 - x_2 + 2x_3 + x_6 &= 4 \\
2x_1 + 3x_2 - x_3 + x_7 &= 2 \\
x_1, x_2, x_3, x_4, x_5, x_6, x_7 &\geq 0
\end{aligned}
$$

```
package Optimization;

public class Test_revisedSimplex extends Object {

  public static void main(String args[]) {

    int n = 7;
    int m = 4;
    double eps = 1.0e-5;
    double a[][] = {{0,  5,  5,  3,  0,  0,  0, 0},
                    {3,  1,  3,  1,  1,  0,  0, 0},
                    {2, -1,  0,  3,  0,  1,  0, 0},
                    {4,  2, -1,  2,  0,  0,  1, 0},
                    {2,  2,  3, -1,  0,  0,  0, 1}};
    int basicvar[] = new int[m + 3];

    Optimize.revisedSimplex(true, n, m, a, eps, basicvar);
```

```
    if (basicvar[m+1] > 0)
      System.out.println("No feasible solution.");
    else {
      if (basicvar[m+2] > 0)
        System.out.println("Objective function is unbound.");
      else {
        System.out.println("Optimal solution found." +
                           "\n\n Basic variable   Value");
        for (int i=1; i<=m; i++)
          System.out.printf(" %6d %17.5f\n", basicvar[i], a[i][0]);
        System.out.println("\nOptimal value of the objective function = " +
                           a[0][0]);
      }
    }
  }
}
```

Output:

```
Optimal solution found.

 Basic variable   Value
      3           1.03448
      1           1.10345
      4           0.03448
      2           0.27586

Optimal value of the objective function = 10.0
```

10.2 Dual Simplex Method

The following is a *linear programming problem* in standard form:

$$\text{maximize} \quad \sum_{j=1}^{n} c_j x_j$$

$$\text{subject to} \quad \sum_{j=1}^{n} a_{ij} x_j \le b_i \qquad (i = 1,2,\ldots,m)$$

$$x_j \ge 0 \qquad (j = 1,2,\ldots,n)$$

After introducing the slack variables x_{n+1}, x_{n+2}, ..., x_{n+m}, the problem can be written as:

$$\text{maximize} \quad \sum_{j=1}^{n} c_j x_j$$

$$\text{subject to} \quad \sum_{j=1}^{n} a_{ij} x_j = b_i \qquad (i = 1,2,\ldots,m)$$

$$x_j \geq 0 \qquad (j = 1,2,\ldots,n)$$

or, in matrix notation:

maximize cx
subject to Ax = b
$x \geq 0$

The matrix A has m rows and n+m columns with the last m columns forming an identity matrix. The vector x is of length n+m, and the column b is of length m. The *dual problem* of the linear programming problem in standard form is:

$$\text{minimize} \quad \sum_{i=1}^{m} b_i y_i$$

$$\text{subject to} \quad \sum_{i=1}^{m} a_{ij} y_i \geq c_j \qquad (j = m+1, m+2, \ldots, m+n)$$

$$y_i \geq 0 \qquad (i = 1,2,\ldots,m)$$

A dual variable y_k is basic if and only if the corresponding primal variable x_k is nonbasic. The linear programming problem can be solved by the *dual simplex method* [L54]. A basic feasible solution x^* partitions x into x_B (m basic variables) and x_N (n nonbasic variables). This corresponds to the partition of matrix A into A_B and A_N. Each iteration of the dual simplex method [C83, SDK83] can be described as follows:

Step 1. If $x_B^* \geq 0$ then stop (x^* is an optimal solution), otherwise choose any basic variable x_i with $x_i^* < 0$ as the leaving variable.

Step 2. Solve the system $uA_B = e_k$ where e_k is the k-th row of the identity matrix such that x_i appears in the k-th position of the basis. Compute $\gamma_N = uA_N$.

Step 3. Let P be the set of nonbasic variables x_j for which $\gamma_j < 0$. If P is empty then stop (the problem is infeasible), otherwise, find the entering variable x_j where x_j is in P that minimizes c_j/γ_j.

Step 4. Solve the system $A_B\beta = \alpha$ where α is the entering column.

Step 5. Set the value x_j^* of the entering variable to be x_i^*/γ_j. Replace the values x_B^* of the basic variables by $x_B^* - (x_i^*/\gamma_j)\beta$. Replace the leaving column of A_B by the entering column. Replace the leaving variable in the basis by the entering variable. Set $c_i = -c_j/\gamma_j$. Add $c_i\gamma_k$ to each c_k where $k \neq i$.

The following procedure is the dual simplex method that solves the linear programming problem of the form:

$$\text{minimize (or maximize)} \sum_{j=1}^{n} a_{0j} x_j$$

$$\text{subject to} \quad \sum_{j=1}^{n-m} a_{ij} x_j + x_{n-m+1} = a_{i0} \qquad (i = 1,2,\ldots,m)$$

$$x_j \geq 0 \qquad (j = 1,2,\ldots,n)$$

Procedure parameters:

void dualSimplex (*maximize, n, m, a, epsilon, basicvar*)

maximize: boolean;
- entry: *maximize* = true, if the objective function is to be maximized; *maximize* = false, if the objective function is to be minimized.

n: int;
- entry: number of variables (this includes the slack variables).

m: int;
- entry: number of constraints.

a: double[*m*+1][*n*+1];
- entry: the coefficients of the constraints are given by *a[i][j]*, *i*=1,2,...,*m*, *j*=1,2,...,*n*–*m*. Note that the slack variables are not needed in the input.
  The coefficients of the objective function are given by *a[0][j]*, *j*=1,2,...,*n*.
  The right hand side of the constraints is given by *a[i][0]*, *i*=1,2,...,*m*.
  The other elements of the matrix *a* are not required as input.
- exit: *a[0][0]* is the optimal value of the objective function.
  a[i][0] is the optimal value of the basic variable *basicvar[i]*, for i = 1, 2, ..., *m*.

epsilon: double;
- entry: for any real number r, if |r| < epsilon then r is equal to 0.

basicvar: int[*n*+*m*+3];
- exit: if there is no feasible solution then *basicvar[m+1]* > 0, otherwise *basicvar[m+1]* = 0;
  if the problem has no finite solution then *basicvar[m+2]* > 0, otherwise *basicvar[m+2]* = 0;
  basicvar[i] is the basic variable in the optimal solution, for i = 1, 2, ..., *m*.

```
public static void dualSimplex(boolean maximize, int n, int m,
                double a[][], double epsilon, int basicvar[])
{
  int i,j,k,index,nm,q,option;
  int idx1=0,idx2=0,p=0;
  int aux1[] = new int[n - m + 1];
```

```
int aux2[] = new int[n - m + 1];
double thres,diff,amt;
boolean infeasible,unbound,abort,iterate;

option = (maximize ? 1 : -1);
infeasible = false;
unbound = false;
nm = n - m;
a[0][0] = 0.0;
for (i=0; i<=nm; i++) {
  amt = 0.0;
  for (j=1; j<=m; j++)
    amt += a[j][i] * a[0][nm+j];
  a[0][i] = amt - a[0][i];
}
for (i=nm+1; i<=n; i++)
  for (j=1; j<=m; j++)
    a[j][i] = ((i == nm+j) ? 1.0 : 0.0);
i = 0;
while ((!infeasible) && (i < nm)) {
  i++;
  amt = a[0][i];
  infeasible = (Math.abs(amt) > epsilon) && (amt*option < 0);
  if (!infeasible) basicvar[m+i] = i;
}
if (!infeasible) {
  for (i=1; i<=m; i++)
    basicvar[i] = nm + i;
  iterate = true;
  do {
    thres = 0.0;
    abort = true;
    i = 0;
    do {
      i++;
      j = m;
      amt = a[i][0];
      if (amt < -epsilon) {
        iterate = false;
        while ((j < n) && !iterate) {
          j++;
          q = basicvar[j];
          iterate = (a[i][q] < -epsilon);
        }
        if (!iterate)
          unbound = true;
        else {
          abort = false;
          if (amt-thres < -epsilon) {
            thres = amt;
```

```
            p = i;
          }
        }
      }
    } while (iterate && (i < m));
    if (iterate) {
      if (abort) {
        unbound = false;
        iterate = false;
      }
      else {
        thres = Double.MAX_VALUE;
        for (j=1; j<=nm; j++)
          aux2[j] = m + j;
        for (i=0; i<=m; i++)
          if ((i != 1) && (!abort)) {
            k = 0;
            for (j=1; j<=nm; j++)
              aux1[j] = aux2[j];
            index = 1;
            for (j=m+1; j<=n; j++)
              if (j == aux1[index]) {
                index++;
                q = basicvar[j];
                amt = a[p][q];
                if (amt < -epsilon) {
                  amt = Math.abs(a[i][q] / amt);
                  diff = amt - thres;
                  if (Math.abs(diff) < epsilon) {
                    k++;
                    aux2[k] = j;
                    abort = false;
                  }
                  else
                    if (diff < 0.0) {
                      thres = amt;
                      idx1 = j;
                      idx2 = q;
                      aux2[1] = 1;
                      k = 1;
                      for (q=2; q<=nm; q++)
                        aux2[q] = 0;
                      abort = true;
                    }
                }
              }
          }
        thres = 1.0 / a[p][idx2];
        basicvar[idx1] = basicvar[p];
        i = (p == 0 ? 1 : 0);
```

```
          do {
            amt = a[i][idx2] * thres;
            a[i][0] -= a[p][0] * amt;
            for (j=m+1; j<=n; j++) {
              q = basicvar[j];
              a[i][q] -= a[p][q] * amt;
            }
            i += ((i == p-1) ? 2 : 1);
          } while (i <= m);
          for (j=m+1; j<=n; j++) {
            q = basicvar[j];
            a[p][q] *= thres;
          }
          a[p][0] *= thres;
          for (i=0; i<=m; i++)
            a[i][idx2] = ((i == 1) ? 1.0 : 0.0);
          basicvar[p] = idx2;
        }
      }
    } while (iterate);
  }
  basicvar[m+1] = (infeasible ? 1 : 0);
  basicvar[m+2] = (unbound ? 1 : 0);
}
```

Example:

Solve the following linear programming problem [C83] with 7 variables and 3 constraints by the dual simplex method:

$$\text{maximize} \quad -5x_1 - 3x_2 - 3x_3 - 6x_4$$

subject to

$$\begin{aligned}
-6x_1 + x_2 + 2x_3 + 4x_4 + x_5 \qquad\qquad &= 14 \\
3x_1 - 2x_2 - x_3 - 5x_4 + \qquad x_6 \qquad &= -25 \\
-2x_1 + x_2 + \qquad 2x_4 + \qquad\qquad x_7 &= 14 \\
x_1, x_2, x_3, x_4, x_5, x_6, x_7 &\ge 0
\end{aligned}$$

```
package Optimization;

public class Test_dualSimplex extends Object {

  public static void main(String args[]) {

    int n = 7;
    int m = 3;
    double eps = 1.0e-5;
```

```
    double a[][] = {{  0, -5, -3, -3, -6,  0,  0,  0},
                    { 14, -6,  1,  2,  4,  0,  0,  0},
                    {-25,  3, -2, -1, -5,  0,  0,  0},
                    { 14, -2,  1,  0,  2,  0,  0,  0}};
    int basicvar[] = new int[n + m + 1];

    Optimize.dualSimplex(true, n, m, a, eps, basicvar);
    if (basicvar[m+1] > 0)
      System.out.println("No feasible solution.");
    else {
      if (basicvar[m+2] > 0)
        System.out.println("Objective function is unbound.");
      else {
        System.out.println("Optimal solution found." +
                           "\n\n Basic variable   Value");
        for (int i=1; i<=m; i++)
          System.out.printf(" %6d %17.4f\n", basicvar[i], a[i][0]);
        System.out.println("\nOptimal value of the objective function = " +
                           a[0][0]);
      }
    }
  }
}
```

Output:

```
Optimal solution found.

 Basic variable   Value
      2           10.0000
      4            1.0000
      7            2.0000

Optimal value of the objective function = -36.0
```

11. Integer Programming

11.1 Zero-One Integer Programming

Solve the *zero-one integer programming problem* of the form:

$$\text{minimize (or maximize)} \quad \sum_{j=1}^{n} c_j x_j$$

$$\text{subject to} \quad \sum_{j=1}^{n} a_{ij} x_j \le b_i \qquad (i = 1,2,\ldots,m)$$

$$x_j = 0 \text{ or } 1, \qquad (j = 1,2,\ldots,n)$$

The problem can be solved by enumerating implicitly all 2^n zero-one vectors of x. The enumeration can be summarized as follows:

a. Choose a free variable x_j and fix it at the value 1.
b. Enumerate each of the completions of the partial solution.
c. The variable x_j is fixed at the value 0, and the process is repeated for the subproblem with x_j equal to 0.

The following procedure [B65, SDK83] uses an effective branching test in the selection of the free variables. The branching test essentially chooses the free variable x_j to be set to 1 in such a way that the sum of the absolute values of the amount by which all constraints are violated is reduced by the most.

Procedure parameters:

void zeroOneIntegerProgramming (*minimize, n, m, a, b, c, sol*)

minimize: boolean;
entry: *minimize* = true, if the objective function is to be minimized;
minimize = false, if the objective function is to be maximized.

n: int;
entry: number of variables.

m: int;
entry: number of constraints.

a: double[m+1][n+1];
entry: the coefficients of the constraints are given by *a[i][j]*, i=1,2,...,m, j=1,2,...,n.
The other elements of the matrix *a* are not required as input.
exit: if an optimal solution is found then *a[0][0]* = 0;
if there is no feasible solution then *a[0][0]* > 0.

b: int[m+1];
entry: *b[j]* is the right hand side of constraint *j*, j=1,2,...,m.

c: int[n+1];
entry: *c[i]* is the coefficent of varible *i* in the objective function, i=1,2,...,n.

sol: int[n+1];
exit: *sol[i]* is the optimal solution of variable *i*, i = 1, 2, ..., n.
sol[0] is the optimal value of the objective function.

```
public static void zeroOneIntegerProgramming(boolean minimize, int n,
                         int m, int a[][], int b[], int c[], int sol[])
{
  int i,j,k,optvalue,elm1=0,elm2,elm3,elm4,idx,sub1,sub2,sub3;
  int item1,item2,item3;
  int ccopy[] = new int[n + 1];
  int aux1[] = new int[n + 1];
  int aux2[] = new int[n + 1];
  int aux3[] = new int[n + 1];
  int aux4[] = new int[n + 2];
  int aux5[] = new int[m + 1];
  int aux6[] = new int[m + 1];
  int aux7[] = new int[m + 1];
  boolean cminus[] = new boolean[n + 1];
  boolean optimalfound,backtrack=false,outer;

// scan for the negative objective coefficients
  if (!minimize)
    for (j=1; j<=n; j++)
      c[j] = -c[j];
  for (j=1; j<=n; j++) {
    cminus[j] = false;
    ccopy[j] = c[j];
  }
  for (j=1; j<=n; j++)
    if (c[j] < 0) {
      cminus[j] = true;
      c[j] = -c[j];
      for (i=1; i<=m; i++) {
        b[i] -= a[i][j];
        a[i][j] = -a[i][j];
      }
    }

  for (i=1; i<=m; i++)
    aux5[i] = b[i];
  elm4 = 1;
  for (j=1; j<=n; j++) {
    aux3[j] = 0;
    elm4 += c[j];
  }
  optvalue = elm4 + elm4;
  sub2 = 0;
  sub3 = 0;
  elm4 = 0;
  aux4[1] = 0;
  optimalfound = false;
  iterate:
```

```
  while (true) {
    if (backtrack) {
//    backtracking
      backtrack = false;
      outer = false;
      for (j=1; j<=n; j++)
        if (aux3[j] < 0) aux3[j] = 0;
      if (sub2 > 0)
        do {
          sub1 = sub3;
          sub3 -= aux4[sub2+1];
          for (j=sub3+1; j<=sub1; j++)
            aux3[aux2[j]] = 0;
          sub1 = Math.abs(aux1[sub2]);
          aux4[sub2] += sub1;
          for (j=sub3-sub1+1; j<=sub3; j++) {
            sub1 = aux2[j];
            aux3[sub1] = 2;
            elm4 -= c[sub1];
            for (i=1; i<=m; i++)
              aux5[i] += a[i][sub1];
          }
          sub2--;
          if (aux1[sub2+1] >= 0) {
            outer = true;
            continue iterate;
          }
        } while (sub2 != 0);
      if (outer) continue;
      sol[0] = optvalue;
      a[0][0] = (optimalfound ? 0 : 1);
      for (j=1; j<=n; j++)
        if (cminus[j]) {
          sol[j] = ((sol[j] == 0) ? 1 : 0);
          sol[0] += ccopy[j];
        }
      for (j=1; j<=n; j++)
        c[j] = ccopy[j];
      if (!minimize) sol[0] = -sol[0];
      return;
    }
    sub1 = 0;
    idx = 0;
    for (i=1; i<=m; i++) {
      item1 = aux5[i];
      if (item1 < 0) {
//      infeasible constraint i
        sub1++;
        elm3 = 0;
        elm1 = item1;
```

```
      elm2 = -Integer.MAX_VALUE;
      for (j=1; j<=n; j++)
        if (aux3[j] <= 0)
          if (c[j] + elm4 >= optvalue) {
            aux3[j] = 2;
            aux4[sub2+1]++;
            sub3++;
            aux2[sub3] = j;
          }
          else {
            item2 = a[i][j];
            if (item2 < 0) {
              elm1 -= item2;
              elm3 += c[j];
              if (elm2 < item2) elm2 = item2;
            }
          }
      if (elm1 < 0) {
        backtrack = true;
        continue iterate;
      }
      if (elm1 + elm2 < 0) {
        if (elm3 + elm4 >= optvalue) {
          backtrack = true;
          continue iterate;
        }
        for (j=1; j<=n; j++) {
          item2 = a[i][j];
          item3 = aux3[j];
          if (item2 < 0) {
            if (item3 == 0) {
              aux3[j] = -2;
              for (k=1; k<=idx; k++) {
                aux7[k] -= a[aux6[k]][j];
                if (aux7[k] < 0) {
                  backtrack = true;
                  continue iterate;
                }
              }
            }
          }
          else
            if (item3 < 0) {
              elm1 -= item2;
              if (elm1 < 0) {
                backtrack = true;
                continue iterate;
              }
              elm3 += c[j];
              if (elm3 + elm4 >= optvalue) {
```

```
                    backtrack = true;
                    continue iterate;
                  }
                }
              }
              idx++;
              aux6[idx] = i;
              aux7[idx] = elm1;
            }
          }
        }
        if (sub1 == 0) {
//        updating the best solution
          optvalue = elm4;
          optimalfound = true;
          for (j=1; j<=n; j++)
            sol[j] = ((aux3[j] == 1) ? 1 : 0);
          backtrack = true;
          continue iterate;
        }
        if (idx == 0) {
          sub1 = 0;
          elm3 = -Integer.MAX_VALUE;
          for (j=1; j<=n; j++)
            if (aux3[j] == 0) {
              elm2 = 0;
              for (i=1; i<=m; i++) {
                item1 = aux5[i];
                item2 = a[i][j];
                if (item1 < item2) elm2 += (item1 - item2);
              }
              item1 = c[j];
              if ((elm2 > elm3) || (elm2 == elm3) && (item1 < elm1)) {
                elm1 = item1;
                elm3 = elm2;
                sub1 = j;
              }
            }
          if (sub1 == 0) {
            backtrack = true;
            continue iterate;
          }
          sub2++;
          aux4[sub2+1] = 0;
          sub3++;
          aux2[sub3] = sub1;
          aux1[sub2] = 1;
          aux3[sub1] = 1;
          elm4 += c[sub1];
          for (i=1; i<=m; i++)
```

```
          aux5[i] -= a[i][sub1];
      }
      else {
        sub2++;
        aux1[sub2] = 0;
        aux4[sub2+1] = 0;
        for (j=1; j<=n; j++)
          if (aux3[j] < 0) {
            sub3++;
            aux2[sub3] = j;
            aux1[sub2]--;
            elm4 += c[j];
            aux3[j] = 1;
            for (i=1; i<=m; i++)
              aux5[i] -= a[i][j];
          }
      }
    }
}
```

Example:

Solve the following zero-one integer programming problem with 4 variables and 3 constraints:

$$\begin{aligned} \text{minimize} \quad & 12x_1 + 14x_2 + 23x_3 + 36x_4 \\ \text{subject to} \quad & \\ & -10x_1 - 13x_2 - 11x_3 - 23x_4 \le -10 \\ & -4x_1 - 6x_2 - 11x_3 - 16x_4 \le -12 \\ & -12x_1 - 10x_2 - 5x_3 - 9x_4 \le -8 \\ & x_1, x_2, x_3, x_4 = 0 \text{ or } 1 \end{aligned}$$

```
package Optimization;

public class Test_zeroOneProgram extends Object {

  public static void main(String args[]) {

    int n = 4;
    int m = 3;
    int a[][] = {{0,   0,   0,   0,   0},
                 {0, -10, -13, -11, -23},
                 {0,  -4,  -6, -11, -16},
                 {0, -12, -10,  -5,  -9}};
    int b[] = {0, -10, -12, -8};
    int c[] = {0,  12,  14,  23, 36};
    int sol[] = new int[n + 1];
```

```
    Optimize.zeroOneIntegerProgramming(true, n, m, a, b, c, sol);
    if (a[0][0] > 0)
      System.out.println("No feasible solution.");
    else {
      System.out.print("Optimal solution found.\n Solution vector: ");
      for (int i=1; i<=n; i++)
        System.out.print("  " + sol[i]);
      System.out.println("\n Optimal value of the objective function = " +
                         sol[0]);
    }
  }
}
```

Output:

```
Optimal solution found.
 Solution vector:   1  0  1  0
 Optimal value of the objective function = 35
```

11.2 All Integer Programming

Solve the following *integer linear programming problem* of the form:

$$\text{minimize} \quad \sum_{j=1}^{n} a_{0j} x_j$$

$$\text{subject to} \quad \sum_{j=1}^{n} a_{ij} x_j \le a_{i,n+1} \qquad (i = 1,2,\ldots,m)$$

$$x_j \ge 0 \text{ integer} \qquad (j = 1,2,\ldots,n)$$

Let Q be the set of nonbasic variable indices. In vector notation, the problem can be written as:

$$\text{minimize} \quad x_0$$

$$\text{subject to} \quad x = d_0 + \sum_{j \in Q} d_j(-x_j)$$

$$x_i \ge 0 \text{ integer} \qquad (i = 1,2,\ldots,n)$$

It is assumed that the vectors d_j, j=m+1,...,n are lexicographically negative. The Gomory cutting plane technique [G60, G65, SDK83] solves this problem by starting with an integer tableau and a lexicographic dual feasible solution:

Step 1. Select a cut-generating row $d_{k0} < 0$, $k \neq 0$. If no such row exists then the current solution is optimal.
Step 2. Among the $d_{kj} < 0$, select the pivot column that is lexicographically largest. If no such column is found then no integer feasible solution exists.
Step 3. A cut inequality from the row k that is not satisfied at the current primal solution is derived. This new row is appended at the bottom of the tableau, and is used as a pivot in the current solution.
Step 4. Perform a dual simplex pivoting operation.
Step 5. Remove the appended row and go back to Step 1.

Procedure parameters:

void integerProgramming (*n, m, a*)

n: int;
entry: number of variables.

m: int;
entry: number of constraints.

a: double[*n*+*m*][*n*+1];
entry: the coefficients of the objective function are given by *a[0][j]*, *j*=1,2,...,*n*;
the coefficients of the constraints are given by *a[i][j]*, *i*=1,2,...,*m*, *j*=1,2,...,*n*;
the right hand side of the constraints is given by *a[i][n+1]*, *i*=1,2,...,*m*.
It is required that each of the j columns, j=1,2,...,*n*, of matrix *a* is lexicographically positive.
The other elements of the matrix *a* are not required as input.
exit: if an optimal solution is found then *a[0][n+1]* = 0;
if there is no feasible solution then *a[0][n+1]* > 0.
The optimal value of the objective function is given by *a[0][0]*.
The optimal solution vector is given by *a[m+i][0]*, *i*=1,2,...,*n*.

```
public static void integerProgramming(int n, int m, int a[][])
{
  int i,j,k,l,np,num,r,r1,s,t,count,c,denom,temp,p;
  boolean b,iter,nofeas=false;

  for (j=1; j<=n; j++)
    a[m+j][j] = -1;
  m += n;
  count = 0;
  np = n+1;
  do {
    count++;
    r = 0;
    do {
      r++;
```

```
      iter = a[r][np] < 0;
    } while (!iter && (r != m));
    if (iter) {
      k = 0;
      do {
        k++;
        iter = a[r][k] < 0;
      } while (!iter && (k != n));
      nofeas = !iter;
      if (iter) {
        l = k;
        for (j=k+1; j<=n; j++)
          if (a[r][j] < 0) {
            i = -1;
            do {
              i++;
              s = a[i][j] - a[i][l];
            } while (s == 0);
            if (s < 0) l = j;
          }
        s = 0;
        while (a[s][l] == 0)
          s++;
        num = -a[r][l];
        denom = 1;
        for (j=1; j<=n; j++)
          if ((a[r][j] < 0) && (j != l)) {
            i = s - 1;
            b = true;
            while (b && (i >= 0)) {
              b = (a[i][j] == 0);
              i--;
            }
            if (b) {
              i = a[s][j];
              r1 = a[s][l];
              temp = i / r1;
              if (temp*r1 > i) temp--;
              if ((temp+1)*r1 <= i) temp++;
              t = temp;
              if ((t*r1 == i) && (t > 1)) {
                i = s;
                do {
                  i++;
                  r1 = t*a[i][l] - a[i][j];
                } while (r1 == 0);
                if (r1 > 0) t--;
              }
              c = -a[r][j];
              if (c*denom > t*num) {
```

```
                num = c;
                denom = t;
              }
            }
          }
        for (j=1; j<=np; j++)
          if (j != l) {
            p = a[r][j] * denom;
            temp = p / num;
            if (temp*num > p) temp--;
            if ((temp+1)*num <= p) temp++;
            c = temp;
            if (c != 0)
              for (i=0; i<=m; i++)
                a[i][j] += c * a[i][l];
          }
      }
    }
  } while (iter && !nofeas);
  a[0][0] = -a[0][n+1];
  a[0][n+1] = (nofeas ? 1 : 0);
  for (j=1; j<=n; j++)
    a[m-n+j][0] = a[m-n+j][n+1];
}
```

Example:

Solve the following integer programming problem with 3 variables and 2 constraints:

$$
\begin{aligned}
&\text{minimize} \quad 3x_1 + 3x_2 + 4x_3 \\
&\text{subject to} \\
&\qquad -2x_1 - 2x_2 - 3x_3 \le -12 \\
&\qquad -4x_1 - x_2 - x_3 \le -10 \\
&\qquad\qquad x_1, x_2, x_3 \ge 0 \text{ integer}
\end{aligned}
$$

```
package Optimization;

public class Test_integerProgram extends Object {

  public static void main(String args[]) {
    int n = 3;
    int m = 2;
    int a[][] = {{0,  3,  3,  4,   0},
                 {0, -2, -2, -3, -12},
                 {0, -4, -1, -1, -10},
                 {0,  0,  0,  0,   0},
                 {0,  0,  0,  0,   0},
```

```
                {0,  0,  0,  0,   0}};

    Optimize.integerProgramming(n, m, a);
    if (a[0][n+1] > 0)
      System.out.println("No feasible solution.");
    else {
      System.out.print("Optimal solution found.\n Solution: ");
      for (int i=1; i<=n; i++)
        System.out.print("  " + a[m+i][0]);
      System.out.println("\n Optimal value of the objective function = " +
                         a[0][0]);
    }
  }
}
```

Output:

```
Optimal solution found.
 Solution:  3  0  2
 Optimal value of the objective function = 17
```

11.3 Mixed Integer Programming

In a *mixed integer programming problem* some of the variables are restricted to integer values, but the rest are ordinary continuous variables. It has the form:

$$\text{minimize} \quad \sum_{j=1}^{n} c_j x_j$$

$$\sum_{j=1}^{n} a_{ij} x_j \le, =, \ge b_i \qquad (i = 1, 2, \ldots, m)$$

$$\text{subject to} \quad x_j \ge 0 \qquad (j = 1, 2, \ldots, n)$$

$$x_j \text{ is an integer,} \quad (j = 1, 2, \ldots, k, \quad \text{where } k \le n)$$

If k = 0, this becomes a linear programming problem. If k = n, this becomes a pure integer programming problem.

The following branch-and-bound procedure [LD60, KM73] solves the mixed integer programming problem by imbedding a dual simplex algorithm in the program to obtain the initial continuous solution. Each integer trial is evaluated. The integer variables are tested one at a time.

Step 1. Relax the integrality restrictions and find a continuous solution of the linear programming problem. If the continuous solution satisfies all the integrality restrictions, then the optimal solution of the original mixed integer programming problem has been found, otherwise some integrality restriction is violated.

Step 2. Choose an integral variable x_j that is currently fractional with value p. Let q be the integral part of p (q is the largest integer less than or equal to p). Create two branch problems: one of these two problems has the added restriction $x_j \le q$, and the other problem has the added restriction $x_j \ge q+1$. This procedure is repeated.

An upper bound and a lower bound are generated in the branching process. Feasible integral solutions are used as upper bounds. Lower bounds are obtained by taking the smallest optimal objective value for a linear programming relaxation among all current active nodes of the enumeration tree.

Procedure parameters:

void mixedIntegerProgram (*numvar, numconstraints, numintvar, upbound, constrainttype, tableau*)

numvar: int;
entry: number of variables, corresponding to n in the above problem description.

numconstraints: int;
entry: number of constraints, corresponding to m in the above problem description.

numintvar: int;
entry: number of integer variables, corresponding to k in the above problem description, $0 \le numintvar \le numvar$.

upbound: double[*numvar*+1];
entry: *upbound[i]* is the upper bound of the i-th integer variable, for i=1,2,..., *numintvar*.
upbound[j] should be set to zero for all the continuous variables, j = *numintvar*+1,...,n.

constrainttype: int[m+2];
entry: *constrainttype[i]* = *−1* if the i-th constraint is of the type $a_{ij}x_j \le b_i$,
constrainttype[i] = *0* if the i-th constraint is of the type $a_{ij}x_j = b_i$,
constrainttype[i] = *+1* if the i-th constraint is of the type $a_{ij}x_j \ge b_i$,
for i = 2,3,..., *numconstraints*+1.
exit: *constrainttype[0]* = 0 if an optimal solution is found,
constrainttype[0] = 1 if the problem is not feasible.

tableau: double[*numconstraints*+3][*numvar*+2];
entry: The coefficients of the objective function are stored in row 1 of this array, i.e., *tableau[1][j+1]* = c_j for j = 1,2,..., *numvar*.
The right hand side of the i-th constraint is stored in column 1 of the array, i.e., *tableau[i][1]* = b_i for i = 2,3,..., *numconstraints*+1.
The left hand side coefficients of the i-th constraint are stored in row i+1 of the array, i.e., *tableau[i+1][j+1]* = a_{ij},
for j = 1,2,..., *numvar*, and i = 1,2,..., *numconstraints*.
The values of all other entries of *tableau* do not need to be set.
exit: If an optimal solution is found, i.e., the return value of *constrainttype[0]* is zero, then the optimal value of the objective function is stored in *tableau[0][0]*, and the value of the variable x_j is

stored in *tableau[0][j]* for $j = 1, 2, \ldots, numvar$.

```
public static void mixedIntegerProgram(int numvar, int numconstraints,
                   int numintvar, double upbound[], int constrainttype[],
                   double tableau[][])
{
  int n = numvar + 1;
  int m = numconstraints + 1;

  int controlb,controld,i,j,idx1,idx5,idx9,nminus1,loopcount;
  int p,q,r,numiterations,mille;
  int controla=0,controlc=0,controle=0,pivotcol=0;
  int k=0,idx2=0,idx3=0,idx4=0,idx6=0,idx7=0,idx8=0,pivotrow=0;
  int aux1[] = new int[n+1];
  int aux2[] = new int[n+1];
  int aux3[] = new int[n+1];
  int aux4[] = new int[m+1];
  int aux5[] = new int[numintvar+1];
  int aux6[] = new int[numintvar+1];
  int aux7[] = new int[numintvar+1];
  int aux8[] = new int[numintvar+2];
  int aux9[][] = new int[m+1][numintvar+1];
  double objestimate,objvaltolerance,vala,vald,vale,valf,valg;
  double valj,valh,quan1,ajw,quan4,amt1,amt2,amt3;
  double quan5,quan6,quan7,quan8,tmp,quan9;
  double vali=0.,threshold=0.,quan3=0.,quan2=0.,valb=0.,valc=0.;
  double large=1.e34,verylarge=1.e35;
  double origobj[] = new double[n];
  double wk1[] = new double[n + 1];
  double wk2[] = new double[numintvar+2];
  double wk3[] = new double[numintvar+2];
  double wk4[] = new double[numintvar+2];
  double wk5[][] = new double[m+1][n+1];
  double wk6[][] = new double[m+2][n*(n+1)/2+1];
  boolean tempbest=false,skip=false;
  boolean skipa=false,skipb=false,skipc=false,skipd=false,skipe=false;
  boolean overa=false,overb=false,overc=false,prow=false;

  quan8 = 1.0;
  controlb = 1;
  aux8[1] = 1;
  controld = 1;
  numiterations = 0;
  mille = 1000;
  idx5 = mille;
  quan9 = 0.00001;

  for (i=0; i<=n; i++) {
    tableau[0][i] = 0.;
```

```
  tableau[m+1][i] = 0.;
}
tableau[1][1] = 0.;
nminus1 = n - 1;
for (i=1; i<=nminus1; i++)
  origobj[i] = tableau[1][i+1];
for (i=1; i<=n; i++)
  wk1[i] = 0.;
loopcount = 1;
objestimate = verylarge;
objvaltolerance = 0.1;
constrainttype[1] = 0;

if (m >= 2) {
  for (i=2; i<=m; i++)
    if (constrainttype[i] < 0)
      tableau[i][1] = -tableau[i][1];
    else
      for (j=2; j<=n; j++)
        tableau[i][j] = -tableau[i][j];
}
for (i=2; i<=n; i++)
  if (upbound[i-1] <= 0) upbound[i-1] = (double) mille;
// set solution vector of to zero and save original upper bounds
for (i=2; i<=n; i++)
  aux2[i-1] = 0;
// initialize row and column identifiers and slacks
if (m >= 2)
  for (i=2; i<=m; i++)
    if (constrainttype[i] != 0) constrainttype[i] = 1 - i;
vala = tableau[1][1];
aux1[1] = 0;
for (j=2; j<=n; j++)
  if (tableau[1][j] < 0.) {
    for (i=1; i<=m; i++) {
      tableau[i][1] += tableau[i][j] * upbound[j-1];
      tableau[i][j] = -tableau[i][j];
    }
    aux1[j] = mille + j - 1;
  }
  else
    aux1[j] = j - 1;
// finish initializing the tableau
iterate:
while (true) {
  // reverse sign is column of zero slack row is negative
  if (!tempbest) {
    if (!skipd) {
      if (m >= 2) {
        for (k=2; k<=m; k++)
```

```
        if (constrainttype[k] == 0) {
          if (tableau[k][1] < 0.)
            for (p=1; p<=n; p++)
              tableau[k][p] = -tableau[k][p];
        }
    }
    // next pivot step, start dual lp, choose pivot row
    valc = 0.0;
    if (m >= 2) {
      for (i=2; i<=m; i++)
        if (tableau[i][1] > 0.)
          if (tableau[i][1] > valc) {
            valc = tableau[i][1];
            pivotrow = i;
          }
    }
    // if no positive value then linear programming is finished,
    // primal feasible
    if (valc <= 0.) {
      tempbest = true;
      prow = true;
      continue iterate;
    }
  }
  skipd = false;
  // choose pivot column
  valc = -verylarge;
  skip = false;
  if (n >= 2) {
    pivotcol = 0;
    for (j=2; j<=n; j++)
      if (tableau[pivotrow][j] < 0.) {
        valf = tableau[1][j] / tableau[pivotrow][j];
        if (valf > valc) {
          valc = valf;
          pivotcol = j;
        }
        else
          if (valf == valc)
            if (tableau[pivotrow][j] < tableau[pivotrow][pivotcol])
              pivotcol = j;
      }
    if (pivotcol != 0) skip = true;
  }
  if (!skip) {
    switch (controld) {
      case 1:
        constrainttype[0] = 1;
        return;
      case 2:
```

```
        if (vali == 0.)
          wk3[idx6] = -1.;
        else
          wk4[idx6] = (double) mille;
        overc = true;
        break;
      case 3:
        if (controla == 1)
          wk4[idx6] = (double) mille;
        else
          if (controla == 2)
            wk3[idx6] = -1.;
          else {
            if ((wk4[idx6] - wk2[idx6] - 1.) == 0.)
              wk4[idx6] = (double) mille;
            else {
              wk3[idx6] = -1.;
            }
          }
        overc = true;
        break;
      case 4:
        idx6--;
        overb = true;
        break;
      case 5:
        if (controlc == 1) wk4[idx6] = (double) mille;
        overc = true;
        break;
    }
  }
  skip = false;
  if (!overa && !overb && !overc) {
    // pivot step
    valj = tableau[pivotrow][pivotcol];
    // update tableau
    for (j=1; j<=n; j++)
      if (tableau[pivotrow][j] != 0.) {
        if (j != pivotcol) {
          valh = tableau[pivotrow][j] / valj;
          for (i=1; i<=m; i++)
            if (tableau[i][pivotcol] != 0.) {
              if (i != pivotrow) {
                tableau[i][j] -= valh * tableau[i][pivotcol];
                if (Math.abs(tableau[i][j]) <= quan9) tableau[i][j] = 0.;
              }
            }
        }
      }
    for (j=1; j<=n; j++)
```

```
          tableau[pivotrow][j] /= valj;
        // exchange row and column identifiers
        idx1 = constrainttype[pivotrow];
        constrainttype[pivotrow] = aux1[pivotcol];
        if (idx1 == 0) {
        // if pivot row is zero slack, set modifiers pivot column zero.
          for (i=1; i<=m; i++)
            tableau[i][pivotcol] = tableau[i][n];
          aux1[pivotcol] = aux1[n];
          n--;
        }
        else {
          for (i=1; i<=m; i++)
            tableau[i][pivotcol] = -tableau[i][pivotcol] / valj;
          aux1[pivotcol] = idx1;
          tableau[pivotrow][pivotcol] = 1. / valj;
        }
        // count the number of iterations
        numiterations++;
        if ((constrainttype[pivotrow] + mille) == 0) {
          for (j=1; j<=n; j++)
            tableau[pivotrow][j] = tableau[m][j];
          constrainttype[pivotrow] = constrainttype[m];
          m--;
        }
      }
    }
    if (!prow) {
      if (!overa && !overb && !overc) {
        tempbest = false;
        switch (controld) {
          case 1:
            continue iterate;
          case 2:
            if (tableau[1][1] < threshold) continue iterate;
            if (vali == 0.)
              wk3[idx6] = -1.;
            else
              wk4[idx6] = (double) mille;
            overc = true;
            break;
          case 3:
            if (tableau[1][1] < threshold) continue iterate;
            if (controla == 1)
              wk4[idx6] = (double) mille;
            else
              if (controla == 2)
                wk3[idx6] = -1.;
              else {
                if ((wk4[idx6] - wk2[idx6] - 1.) == 0.)
```

```
              wk4[idx6] = (double) mille;
            else {
              wk3[idx6] = -1.;
            }
          }
          overc = true;
          break;
        case 4:
          if (tableau[1][1] < threshold) continue iterate;
          idx6--;
          overb = true;
          break;
        case 5:
          if (tableau[1][1] < threshold) continue iterate;
          if (controlc == 1) wk4[idx6] = (double) mille;
          overc = true;
          break;
      }
    }
  }
  prow = false;
  tempbest = false;
  if (!overa && !overb && !overc) {
    // if a basis variable exceed its upper bound,
    // pivot on the corresponding row
    if (m >= 2) {
      for (i=2; i<=m; i++)
        if (constrainttype[i] > 0) {
          j = constrainttype[i];
          if (j > mille) j -= mille;
          if ((upbound[j] + tableau[i][1]) < 0.) {
            if ((quan9 + upbound[j] + tableau[i][1]) < 0.) {
              tableau[i][1] = -tableau[i][1] - upbound[j];
              for (k=2; k<=n; k++)
                tableau[i][k] = -tableau[i][k];
              pivotrow = i;
              constrainttype[i] =
                   (j != constrainttype[i]) ? j : (constrainttype[i] + mille);
              skipd = true;
              continue iterate;
            }
            else
              tableau[i][1] = -upbound[j];
          }
        }
    }
    // end of linear programming
    if (m >= 2) {
      for (i=2; i<=m; i++)
        if (constrainttype[i] > 0) {
```

```
          if (constrainttype[i] > mille) {
            j = constrainttype[i] - mille;
            wk1[j] = upbound[j] + tableau[i][1];
          }
          else {
            j = constrainttype[i];
            wk1[j] = -tableau[i][1];
          }
        }
      }
      // set solution vector values for non-basic variables
      for (i=2; i<=n; i++)
        if (aux1[i] > 0) {
          if (aux1[i] > mille) {
            j = aux1[i] - mille;
            wk1[j] = upbound[j];
          }
          else {
            j = aux1[i];
            wk1[j] = 0.;
          }
        }
      skipe = false;
      switch (controld) {
        case 1:
          break;
        case 2:
          overa = true;
          break;
        case 3:
          overa = true;
          break;
        case 4:
          controlb = 2;
          skipe = true;
          break;
        case 5:
          n = idx8;
          for (i=1; i<=m; i++) {
            constrainttype[i] = aux4[i];
            for (j=1; j<=n; j++)
              tableau[i][j] = wk5[i][j];
          }
          for (j=1; j<=n; j++)
            aux1[j] = aux3[j];
          overa = true;
          break;
      }
    }
    if (!overa && !overb && !overc) {
```

```
    if (!skipe) {
      // Continuous solution complete
      // compute absolute tolerance
      valb = tableau[1][1];
      vala = Math.abs(vala - tableau[1][1]);
      if (objvaltolerance > 0.)
        threshold = objvaltolerance * vala + valb;
      else
        if (objvaltolerance == 0.) threshold = verylarge;
      // determine whether continuous solution is mixed integer solution
      skip = false;
      if (m >= 2) {
        for (i=2; i<=m; i++)
          if (constrainttype[i] > 0) {
            if (constrainttype[i] <= mille) {
              if (constrainttype[i] > numintvar) continue;
            }
            else {
              if ((constrainttype[i] - mille - numintvar) > 0) continue;
            }
            quan5 = tableau[i][1];
            quan6 = quan9;
            quan7 = quan8;
            tmp = -quan5 - ((int)(-quan5 / quan7)) * quan7;
            if (tmp > quan6) {
              if ((1.0 - tmp - quan6) > 0) {
                skip = true;
                break;
              }
            }
          }
      }
      if (!skip) {
        // either continuous solution is integer solution,
        // or no integer variables are requested
        for (i=1; i<=nminus1; i++)
          tableau[0][i] = wk1[i];
        tmp = 0.;
        for (i=1; i<=nminus1; i++)
          tmp += wk1[i] * origobj[i];
        tableau[0][0] = tmp;
        constrainttype[0] = 0;
        return;
      }
      skip = false;
      idx5 = 0;
    }
  }
  // integer programming start
  iterateIP1:
```

```
while (true) {
  if (!overa && !overb && !overc) {
    if (!skipe)
      idx6 = 1;
  }
  iterateIP2:
  while (true) {
    if (!overa && !overb && !overc) {
      skipe = false;
      valc = -quan8;
      aux8[idx6+1] = aux8[idx6];
      // choose next integer variable to be constrained
      // try nonbasic variables first,
      // choose one with largest shad price
      for (i=2; i<=n; i++)
        if (aux1[i] > 0) {
          if (aux1[i] < mille) {
            if (aux1[i] > numintvar) continue;
          }
          else {
            if ((aux1[i] - mille - numintvar) > 0) continue;
          }
          if (valc < tableau[1][i]) {
            idx3 = i;
            valc = tableau[1][i];
          }
        }
    }
    // if none left, try basic variables
    if (((valc + quan8) != 0.) || overa || overb || overc) {
      if (!overa && !overb && !overc) {
        // variable chosen
        aux2[idx6] = aux1[idx3];
        wk3[idx6] = -1.;
        aux5[idx6] = idx3;
        aux6[idx6] = 0;
        wk2[idx6] = 0.;
        // if objective function value + shadow price exceeds tolerance,
        // indicate upward direction infeasible
        skip = false;
        if ((tableau[1][1] + tableau[1][idx3] - threshold) >= 0.) {
          wk4[idx6] = (double) mille;
          if (idx6 > 1) {
            aux7[idx6] = 0;
            skip = true;
          }
        }
        else {
          wk4[idx6] = 1.;
          if (idx6 != 1) {
```

```
          // save the tableau
          if (idx6 < idx5) skip = true;
        }
      }
      if (!skip) {
        idx9 = aux8[idx6];
        for (j=1; j<=m; j++) {
          aux9[j][idx6] = constrainttype[j];
          for (k=1; k<=n; k++) {
            i = idx9 + k - 1;
            if (j <= 1) wk6[m+1][i] = aux1[k];
            wk6[j][i] = tableau[j][k];
          }
        }
        aux7[idx6] = n;
        aux8[idx6+1] = idx9 + n;
      }
      skip = false;
      aux1[idx3] = aux1[n];
      for (j=1; j<=m; j++)
        tableau[j][idx3] = tableau[j][n];
      n--;
    }
    if (!overb && !overc) {
      overa = false;
      if (idx6 < numintvar) {
        // constrain next integer variable
        idx6++;
        continue iterateIP2;
      }
      // feasible integer solution obtained
      threshold = tableau[1][1];
      objestimate = 1.;
      // store current best mixed integer solution
      for (i=1; i<=numintvar; i++)
        if (aux2[i] != 0) {
          if (aux2[i] <= mille) {
            j = aux2[i];
            wk1[j] = wk2[i];
          }
          else {
            j = aux2[i] - mille;
            wk1[j] = upbound[j] - wk2[i];
          }
        }
      for (i=1; i<=nminus1; i++)
        tableau[0][i] = wk1[i];
      tmp = 0.;
      for (i=1; i<=nminus1; i++)
        tmp += wk1[i] * origobj[i];
```

```
      tableau[0][0] = tmp;
    }
    iteratecur:
    while (true) {
      if (!overb && !overc) {
        if (idx6 <= 0) {
          if (loopcount == 0) {
            // Optimality has been established
            constrainttype[0] = 0;
            return;
          }
          if (objestimate < verylarge) {
            // Optimality has been established
            constrainttype[0] = 0;
            return;
          }
          loopcount++;
          threshold = (double)(loopcount) * objvaltolerance * vala + valb;
          n = aux7[1];
          for (i=1; i<=m; i++) {
            constrainttype[i] = aux9[i][1];
            for (j=1; j<=n; j++)
              tableau[i][j] = wk6[i][j];
          }
          for (k=1; k<=n; k++)
            aux1[k] = (int)wk6[m+1][k];
          continue iterateIP1;
        }
      }
      if (!overb) {
        overc = false;
        k = (aux2[idx6] <= mille) ? aux2[idx6] : aux2[idx6] - mille;
        idx7 = aux5[idx6];
      }
      overb = false;
      if (wk3[idx6] < 0.) {
        if (wk4[idx6] > upbound[k]) {
          idx6--;
          continue iteratecur;
        }
        else {
          //  top end feasible
          controla = 1;
          if (aux6[idx6] == 0) {
            if (idx6 < idx5) {
              controlc = 1;
              if (idx6 != 1) {
                controla = 4;
                idx4 = idx6 - 1;
                idx6 = 1;
```

```
            }
          }
        }
      }
    }
    else
      controla = (wk4[idx6] > upbound[k]) ? 2 : 3;
    while (true) {
      // retrieve saved tableau
      n = aux7[idx6];
      idx9 = aux8[idx6];
      for (p=1; p<=m; p++) {
        constrainttype[p] = aux9[p][idx6];
        for (q=1; q<=n; q++) {
          r = idx9 + q - 1;
          if (p <= 1) aux1[q] = (int)wk6[m+1][r];
          tableau[p][q] = wk6[p][r];
        }
      }
      switch (controla) {
        case 1:
          break;
        case 2:
          wk2[idx6] = wk3[idx6];
          wk3[idx6] -= 1.;
          tableau[idx7][1] += wk2[idx6];
          constrainttype[idx7] = 0;
          if (Math.abs(tableau[idx7][1]) <= quan9) tableau[idx7][1] = 0.;
          controld = 3;
          tempbest = true;
          continue iterate;
        case 3:
          vald = verylarge;
          vale = -verylarge;
          for (p=2; p<=n; p++)
            if (tableau[idx7][p] != 0.) {
              if (tableau[idx7][p] > 0.) {
                valf = tableau[1][p] / tableau[idx7][p];
                if (valf < vald) vald = valf;
              }
              else {
                valg = tableau[1][p] / tableau[idx7][p];
                if (valg > vale) vale = valg;
              }
            }
          if (vald >= verylarge) {
            // bottom end infeasible
            wk3[idx6] = -1.;
            break;
          }
```

```
        if ((vale + verylarge) <= 0.) {
          // top end infeasible
          wk4[idx6] = (double) mille;
          wk2[idx6] = wk3[idx6];
          wk3[idx6] -= 1.;
        }
        else {
          amt2 = Math.abs(vald * (tableau[idx7][1] + wk3[idx6]));
          amt3 = Math.abs(vale * (tableau[idx7][1] + wk4[idx6]));
          if (amt2 <= amt3) {
            wk2[idx6] = wk3[idx6];
            wk3[idx6] -= 1.;
          }
          else
            break;
        }
        tableau[idx7][1] += wk2[idx6];
        constrainttype[idx7] = 0;
        if (Math.abs(tableau[idx7][1]) <= quan9) tableau[idx7][1] = 0.;
        controld = 3;
        tempbest = true;
        continue iterate;
      case 4:
        for (p=1; p<=idx4; p++) {
          q = aux5[p];
          aux1[q] = aux1[n];
          for (j=1; j<=m; j++) {
            if (wk2[p] >= 1.) {
              if (wk2[p] == 1.)
                tableau[j][1] += tableau[j][q];
              else
                tableau[j][1] += wk2[p] * tableau[j][q];
              controlc = 2;
            }
            tableau[j][q] = tableau[j][n];
          }
          n--;
        }
        idx6 = idx4 + 1;
        controla = 1;
    }
    wk2[idx6] = wk4[idx6];
    wk4[idx6] += 1.;
    if (aux6[idx6] != 0) {
      tableau[idx7][1] += wk2[idx6];
      constrainttype[idx7] = 0;
      if (Math.abs(tableau[idx7][1]) <= quan9) tableau[idx7][1] = 0.;
      controld = 3;
      tempbest = true;
      continue iterate;
```

```
          }
          for (p=1; p<=m; p++) {
            tableau[p][1] += wk2[idx6] * tableau[p][idx7];
            if (Math.abs(tableau[p][1]) <= quan9) tableau[p][1] = 0.;
            tableau[p][idx7] = tableau[p][n];
          }
          aux1[idx7] = aux1[n];
          n--;
          if (tableau[1][1] >= threshold) {
            idx6--;
            continue iteratecur;
          }
          if (idx6 < idx5) {
            for (i=1; i<=m; i++) {
              aux4[i] = constrainttype[i];
              for (j=1; j<=n; j++)
                wk5[i][j] = tableau[i][j];
            }
            for (j=1; j<=n; j++)
              aux3[j] = aux1[j];
            idx8 = n;
            controld = 5;
            continue iterate;
          }
          controld = 3;
          tempbest = true;
          continue iterate;
        }
      }
    }
    // choose next integer variable to be constrained
    if (idx6 < idx5)
      idx5 = idx6;
    else
      if (idx6 == idx5) {
        if (controlb == 1) {
          controld = 4;
          continue iterate;
        }
      }
    controlb = 1;
    valc = -quan8;
    skip = false;
    skipa = false;
    skipb = false;
    if (m >= 2) {
      for (idx7=2; idx7<=m; idx7++)
        if (constrainttype[idx7] > 0) {
          if (constrainttype[idx7] < mille) {
            if (constrainttype[idx7] > numintvar) continue;
```

```
        }
        else {
          if ((constrainttype[idx7] - mille - numintvar) > 0) continue;
        }
        vald = verylarge;
        vale = -verylarge;
        quan1 = -tableau[idx7][1] + quan9;
        quan3 = (double)((int)(quan1));
        quan4 = quan3 + 1.;
        if (n <= 1) {
          skip = true;
          break;
        }
        for (p=2; p<=n; p++)
          if (tableau[idx7][p] != 0.) {
            if (tableau[idx7][p] > 0.) {
              valf = tableau[1][p] / tableau[idx7][p];
              if (valf < vald) vald = valf;
            }
            else {
              valg = tableau[1][p] / tableau[idx7][p];
              if (valg > vale) vale = valg;
            }
          }
        if ((vale + large) <= 0.) {
          skipa = true;
          break;
        }
        if (vald >= large) {
          skipb = true;
          break;
        }
        amt2 = Math.abs(vald * (tableau[idx7][1] + quan3));
        amt3 = Math.abs(vale * (tableau[idx7][1] + quan4));
        amt1 = Math.abs(amt2 - amt3);
        if (amt1 > valc) {
          valc = amt1;
          quan2 = quan3;
          idx2 = idx7;
          vali = (amt2 <= amt3) ? 0. : 1.;
        }
      }
  }
  skipc = false;
  if (!skipa) {
    if (!skipb) {
      if (!skip) {
        quan3 = quan2;
        idx7 = idx2;
        wk2[idx6] = quan3 + vali;
```

```
          wk3[idx6] = wk2[idx6] - 1.;
          wk4[idx6] = wk2[idx6] + 1.;
          skipc = true;
        }
        skip = false;
        if (!skipc) {
        // if no. of cols=1 and right hand side=0,
        // then do not proceed to linear programming
          if (Math.abs(tableau[idx7][1] + quan3) > quan9) {
            idx6--;
            overb = true;
            continue iterateIP2;
          }
          wk3[idx6] = -1.;
          wk4[idx6] = (double) mille;
          wk2[idx6] = quan3;
          aux2[idx6] = constrainttype[idx7];
          constrainttype[idx7] = 0;
          overa = true;
          continue iterateIP2;
        }
      }
      if (!skipc) {
        // constraining variable in lower direction infeasible
        wk3[idx6] = -1.;
        if (Math.abs(tableau[idx7][1] + quan3) <= quan9) {
          vali = 0.;
          wk2[idx6] = quan3 + vali;
          wk4[idx6] = wk2[idx6] + 1.;
          skipc = true;
        }
        else {
          wk4[idx6] = quan3 + 2.;
          vali = 1.;
          wk2[idx6] = quan3 + vali;
          skipc = true;
        }
      }
    }
    if (!skipc) {
      // constraining variable in upper direction infeasible
      wk4[idx6] = (double) mille;
      wk3[idx6] = quan3 - 1.;
      vali = 0.;
      wk2[idx6] = quan3 + vali;
    }
    // save the tableau
    skipc = false;
    idx8 = n;
    idx9 = aux8[idx6];
```

```
        for (p=1; p<=m; p++) {
          aux9[p][idx6] = constrainttype[p];
          for (q=1; q<=n; q++) {
            r = idx9 + q - 1;
            if (p <= 1) wk6[m+1][r] = aux1[q];
            wk6[p][r] = tableau[p][q];
          }
        }
        aux7[idx6] = n;
        aux8[idx6+1] = idx9 + n;
        tableau[idx7][1] += wk2[idx6];
        aux5[idx6] = idx7;
        aux2[idx6] = constrainttype[idx7];
        aux6[idx6] = 1;
        constrainttype[idx7] = 0;
        if (Math.abs(tableau[idx7][1]) <= quan9) tableau[idx7][1] = 0.;
        controld = 2;
        // return to carry out linear programming
        continue iterate;
      }
    }
  }
}
```

Example:

Solve the following mixed integer programming problem [S94]:

minimize $24x_1 + 12x_2 + 16x_3 + 4x_4 + 2x_5 + 3x_6$

subject to

$$x_4 + 3x_5 \geq 15$$

$$x_4 + 2x_6 \geq 10$$

$$2x_4 + x_5 \geq 20$$

$$x_4 \leq 15x_1$$

$$x_5 \leq 20x_2$$

$$x_6 \leq 5x_3$$

$$x_1, x_2, x_3 = 0 \text{ or } 1$$

$$x_4, x_5, x_6 \geq 0$$

```
package Optimization;

public class Test_mixedIntegerProgram extends Object {

  public static void main(String args[]) {

    int numvar = 6;
    int numconstraints = 6;
    int numintvar = 3;
    int constrainttype[] = {0, 0, 1, 1, 1, -1, -1, -1};
    double upbound[] = {0.,1.,1.,1.,0.,0.,0.};
    double tableau[][] = {{0.,  0.,  0.,  0.,  0., 0., 0., 0.},
                          {0.,  0., 24., 12., 16., 4., 2., 3.},
                          {0., 15.,  0.,  0.,  0., 1., 3., 0.},
                          {0., 10.,  0.,  0.,  0., 1., 0., 2.},
                          {0., 20.,  0.,  0.,  0., 2., 1., 0.},
                          {0.,  0.,-15.,  0.,  0., 1., 0., 0.},
                          {0.,  0.,  0.,-20.,  0., 0., 1., 0.},
                          {0.,  0.,  0.,  0., -5., 0., 0., 1.},
                          {0.,  0.,  0.,  0.,  0., 0., 0., 0.}};

    Optimize.mixedIntegerProgram(numvar, numconstraints, numintvar,
                              upbound, constrainttype, tableau);
    if (constrainttype[0] > 0)
      System.out.println("No solution found.\nReturn error code = " +
                         constrainttype[0]);
    else {
      System.out.println("Optimal solution found." + "\nSolution vector: ");
      for (int i=1; i<=numvar; i++)
        System.out.printf("%7.2f", tableau[0][i]);
      System.out.printf("\nOptimal objective function value = %6.2f\n",
                        tableau[0][0]);
    }
  }
}
```

Output:

```
Optimal solution found.
Solution vector:
   1.00   1.00   0.00  10.00   1.67   0.00
Optimal objective function value =  79.33
```

12. Quadratic Programming

Consider the *quadratic programming problem* of the form:

$$\text{minimize} \quad \sum_{j=1}^{n} p_j x_j + \frac{1}{2}\sum_{j=1}^{n}\sum_{k=1}^{n} q_{jk} x_j x_k$$

$$\text{subject to} \quad \sum_{j=1}^{n} a_{ij} x_j \le b_i \qquad (i = 1,2,\ldots,m)$$

$$x_j \ge 0 \qquad (j = 1,2,\ldots,n)$$

The algorithm, developed by Wolfe [W59, KM73], is as follows:

Step 1. Find a basic feasible solution such that all state variable values are nonnegative.

Step 2. Separate the objective function into linear and quadratic terms:

$$\sum_{j=1}^{n} p_j x_j + F$$

Decompose the quadratic function F into an n by n matrix by partial derivatives:

$$F = x_j Q x_i$$

where x_j is an n element row vector, x_i is an n element column vector, and

$$Q_{ij} = \frac{1}{2}\frac{\partial F}{\partial x_{ij}}$$

Step 3. Find the minimum of the augmented tableau by a simplex algorithm.

In the following procedure, the *quadratic programming problem* is represented in the form:

minimize d(j) ∗ x(j) + x(i) ∗ c(i,j) ∗ x(j)
subject to a(i,j) ∗ x(j) ≤ b(i) for i=1,2,...,n, j=1,2,...,n.

Procedure parameters:

int quadraticProgramming (*n, m, a, b, c, d, maxiterations, sol*)

quadraticProgramming: int;
exit: the method returns an integer with the following values:
0: optimal solution found.
1: the objective function is unbounded.
2: maximum number of iterations is exceeded.

n: int;
entry: number of variables.

m: int;
entry: number of constraints.

a: double[*m*+1][*n*+1];
entry: the coefficients of the constraints, *a[i][j]* is the coefficient of the j-th variable in the i-th constraint, *i*=1,2,...,*m*, *j*=1,2,...,*n*.
The other elements of the matrix *a* are not required as input.

b: double[*m*+1];
entry: *b[j]* is the right hand side of constraint *j*, *j*=1,2,...,*m*. They must all be nonnegative.

c: double[*n*+1][*n*+1];
entry: the symmetric matrix representing the quadratic coefficients of the objective function. *c[i][j]* is the coefficent of the term $x_i x_j$, *i*=1,2,...,*n*, *j*=1,2,...,*n*.

d: double[*n*+1];
entry: the coefficients of the linear terms in the objective function. *d[j]* is the coefficent of the term x_j, *j*=1,2,...,*n*.

maxiterations: int;
entry: maximum number of iterations allowed.

sol: double[*n*+1];
exit: *sol[i]* is the optimal solution of variable x_j, $j = 1, 2, \ldots, n$.
sol[0] is the optimal value of the objective function.

```
public static int quadraticProgramming(int n, int m, double a[][], double b[],
                        double c[][], double d[], int maxiterations, double sol[])
{
  int mplusn,m2n2,m2n3,m2n3plus1,big,unrestricted;
  int i,j,iterations,temp,column,row=0;
  int index[] = new int[n+m+1];
  double total,candidate,dividend;
  double tableau[][] = new double[n+m+1][2*m+3*n+2];
  double price[] = new double[2*m+3*n+2];
  double change[] = new double[2*m+3*n+2];
  double net[] = new double[2*m+3*n+2];
  double gain[] = new double[2*m+3*n+2];
  double fraction[] = new double[n+m+1];

  for (i=0; i<=n; i++)
    sol[i] = 0.;
  for (i=0; i<=n; i++)
    for (j=0; j<=n; j++)
      if (i != j) c[i][j] /= 2.;
  big = Integer.MAX_VALUE;
  mplusn = m + n;
  m2n2 = m + m + n + n;
  m2n3 = m2n2 + n;
  m2n3plus1 = m2n3 + 1;
  for (i=1; i<=mplusn; i++)
    for (j=1; j<=m2n3plus1; j++)
      tableau[i][j] = 0.;
  for (i=1; i<=m; i++)
    tableau[i][1] = b[i];
  for (i=m+1; i<=mplusn; i++)
```

```
    tableau[i][1] = -d[i-m];
  for (i=1; i<=m; i++)
    for (j=1; j<=n; j++)
      tableau[i][j+1] = a[i][j];
  for (i=1; i<=n; i++)
    for (j=1; j<=n; j++)
      tableau[i+m][j+1] = 2. * c[i][j];
  for (i=m+1; i<=mplusn; i++)
    for (j=n+2; j<=mplusn+1; j++)
      tableau[i][j] = a[j-n-1][i-m];
  for (i=1; i<=mplusn; i++) {
    temp = i + mplusn + n + 1;
    for (j=m2n2+2; j<=m2n3plus1; j++)
      if (j == temp) tableau[i][j] = 1.;
  }
  for (i=m+1; i<=mplusn; i++) {
    temp = i - m + mplusn + 1;
    for (j=mplusn+2; j<=m2n3plus1; j++)
      if (j == temp) tableau[i][j] = -1.;
  }
  for (j=1; j<=m2n3; j++)
    price[j] = 0.;
  for (i=1; i<=m; i++)
    price[n+1+i] = tableau[i][1];
  for (j=m2n2+2; j<=m2n3plus1; j++)
    price[j] = big - 1;
  for (i=1; i<=mplusn; i++)
    index[i] = m2n3 - mplusn + i;
  iterations = 0;
  while (true) {
    // iteration start
    iterations++;
    for (j=1; j<=m2n3plus1; j++)
      gain[j] = 0.;
    for (j=1; j<=m2n3plus1; j++) {
      total = 0.;
      for (i=1; i<=mplusn; i++)
        total += price[index[i]+1] * tableau[i][j];
      gain[j] = total;
      change[j] = price[j] - gain[j];
    }
    // search for the pivot element
    column = 0;
    candidate = 0.;
    // get the variable with largest gain
    for (i=2; i<=m2n3plus1; i++)
      if (change[i] < candidate) {
        candidate = change[i];
        column = i;
      }
```

```
    if (column <= 0) break;
    unrestricted = 0;
    for (i=1; i<=mplusn; i++) {
      if (tableau[i][column] > 0)
        fraction[i] = tableau[i][1] / tableau[i][column];
      else {
        unrestricted++;
        if (unrestricted == mplusn)
          // objective function is unbounded
          return 1;
        else
          fraction[i] = Double.MAX_VALUE;
      }
    }
    // remove limiting variable
    for (i=1; i<=mplusn; i++)
      if (fraction[i] >= 0) {
        if (fraction[i] > big) fraction[i] = big;
        candidate = fraction[i];
        row = i;
        break;
      }
    for (i=1; i<=mplusn; i++)
      if (candidate > fraction[i]) {
        candidate = fraction[i];
        row = i;
      }
    // perform pivoting and introduce new variable
    dividend = tableau[row][column];
    for (j=1; j<=m2n3plus1; j++)
      tableau[row][j] /= dividend;
    for (i=1; i<=mplusn; i++)
      if (i != row) {
        for (j=1; j<=m2n3plus1; j++)
          net[j] = tableau[row][j] * tableau[i][column] /
                                             tableau[row][column];
        for (j=1; j<=m2n3plus1; j++)
          tableau[i][j] -= net[j];
      }
    price[row] = price[column];
    index[row] = column - 1;
    // recompute the price
    for (j=1; j<=m2n2+1; j++)
      price[j] = 0.;
    for (i=1; i<=mplusn; i++) {
      if (index[i] <= mplusn)
        temp = index[i] + mplusn + 1;
      else {
        if (index[i] > m2n2) continue;
        temp = index[i] - (mplusn - 1);
```

```
      }
      price[temp] = tableau[i][1];
    }
    if (iterations >= maxiterations)
      // maximum number of iterations exceeded
      return 2;
  }
  // return the optimal solution
  total = 0.;
  for (i=1; i<=mplusn; i++)
    if (index[i] <= n) total += d[index[i]] * tableau[i][1];
  sol[0] = total;
  total =0.;
  for (i=1; i<=mplusn; i++)
    for (j=1; j<=mplusn; j++) {
      if (index[i] > n) continue;
      if (index[j] > n) continue;
      total += c[index[i]][index[j]] * tableau[i][1] * tableau[j][1];
    }
  sol[0] += total;
  for (i=1; i<=mplusn; i++)
    if ((tableau[i][1] != 0) && (index[i] <= n))
      sol[index[i]] = tableau[i][1];
  return 0;
}
```

Example:

Solve the following quadratic programming problem:

$$\begin{aligned}
\text{minimize} \quad & 3x_1^2 - 3x_1x_2 - 5x_1x_3 + 4x_2^2 + 8x_2x_3 + 7x_3^2 - 5x_1 - 3x_2 - 2x_3 \\
\text{subject to} \quad & \\
& 13x_1 + 7x_2 + 5x_3 \le 67 \\
& x_1 + 2x_2 + x_3 \le 24 \\
& x_1, x_2, x_3 \ge 0
\end{aligned}$$

```
public class Test_quadraticProgramming extends Object {

  public static void main(String args[]) {

    int k;
    int n = 3;
    int m = 2;
    int itmax = 10;
    double a[][] = {{0,  0, 0, 0},
                    {0, 13, 7, 5},
                    {0,  1, 2, 1}};
    double b[] = {0, 67, 24};
    double c[][] = {{0,  0,  0,  0},
                    {0,  3, -3, -5},
                    {0, -3,  4,  8},
                    {0, -5,  8,  7}};
    double d[] = {0, -5, -3, -2};
    double sol[] = new double[n+1];

    k = Optimize.quadraticProgramming(n, m, a, b, c, d, itmax, sol);
    if (k != 0)
      System.out.println("No solution is found, error code = " + k);
    else {
      System.out.printf("Objective function value = %7.2f\n",sol[0]);
      for (int i=1; i<=n; i++)
        System.out.printf("  x(%d) = %6.3f\n",i,sol[i]);
    }
  }
}
```

Output:

```
Objective function value =   -4.63
  x(1) =  1.376
  x(2) =  0.599
  x(3) =  0.292
```

Appendix A: References

[AbS65] M. Abramowitz and I.A. Stegun, *Handbook of Mathematical Functions*, Dover Publications, Inc., New York, 1965.

[B65] E. Balas, "An Additive Algorithm for Solving Linear Programs with Zero-one Variables," *Operations Research*, Vol. 13, 1965, 517-546.

[BK73] C. Bron, J. Kerbosch, "Algorithm 457 – Finding all Cliques of an Undirected Graph," *Communications of the ACM*, Vol. 16, 1973, 575-577.

[B72] J.R. Brown, "Chromatic Scheduling and the Chromatic Number Problem," *Management Science*, Vol. 19, December, Part I, 1972, 456-463.

[BM76] J.A. Bondy, U.S.R. Murty, *Graph Theory with Applications*, Elsevier North-Holland, 1976.

[B57] C. Berge, "Two Theorems in Graph Theory," *Proceedings of the Academy of Sciences of USA*, Vol. 43, 1957, 842-844.

[B99] J. Burkardt, *GRAFPACK*, School of Computational Science, Florida State University, Tallahassee, Flordia, USA, 1999.

[B00] J. Burkardt, *CODEPACK*, School of Computational Science, Florida State University, Tallahassee, Flordia, USA, 2000.

[BD80] R.E. Burkard, U. Derigs, *Assignment and Matching Problems*, Lecture Notes in Economics and Mathematical Systems 184, Springer-Verlag, 1980.

[CT80] G. Carpaneto, P. Toth, "Algorithm 548 –Solution of the Assignment Problem," *ACM Transactions on Mathematical Software*, Vol. 6, No. 1, March 1980, 104-111.

[C83] V. Chvátal, *Linear Programming*, W. H. Freeman and Company, 1983.

[DOH54] G.B. Dantzig, W. Orchard-Hay, "The Product Form for the Inverse in the Simplex Method," *Mathematical Tables and Other Aids to Computation*, Vol. 8, No. 46, April 1954, 64-67.

[D77] H.K. DeWitt, "The Theory of Random Graphs with Applications to the Probabilistic Analysis of Optimization Algorithms," Ph.D. dissertation, Computer Science Department, University of California, Los Angeles, 1977.

[E65] J. Edmonds, "Paths, Trees and Flowers," *Canadian Journal of Mathematics*, Vol. 17, 1965, 449-467.

[EJ73] J. Edmonds, E.L. Johnson, "Matching, Euler Tours and the Chinese Postman," *Mathematical Programming*, Vol. 5, 1973, 88-124.

[E79] S. Even, *Graph Algorithms*, Computer Science Press, 1979.

[ET75] S. Even, R.E. Tarjan, "Network Flow and Testing Graph Connectivity," *SIAM Journal on Computing*, Vol. 4, 1975, 507-518.

[F62] R.W. Floyd, "Algorithm 97 – Shortest Path," *Communications of the ACM*, Vol. 5, 1962, 345.

[F56] L.R. Ford, Jr., "Network Flow Theory," The Rand Corporation, Report P-923, August 1956.

[FF56] L.R. Ford, Jr., D.R. Fulkerson, "Maximal Flow through a Network," *Canadian Journal of Mathematics*, Vol. 8, 1956, 399-404.

[G85] A. Gibbons, *Algorithmic Graph Theory*, Cambridge University Press, 1985.

[G60] R. Gomory, "All-integer Integer Programming," *IBM Research Center, Report RC-189*, 1960.

[G65] R. Gomory, "On the Relation between Integer and Non-integer Solutions to Linear Programs," *Proceedings of the National Academy of Sciences of the United States of America*, Vol. 53, 1965, 260-265.

[H62] F. Harary, "The Maximum Connectivity of a Graph," *Proceedings of the National Academy of Sciences of the United States of America*, Vol. 48, 1962, 1142-1146.

[H69] F. Harary, *Graph Theory*, Addison-Wesley Publishing Company, 1969.

[HT74] J.E. Hopcroft, R.E. Tarjan, "Efficient Planarity Testing," *Journal of the ACM*, Vol. 21, October 1974, 549-568.

[JK91] R. Johnsonbaugh, M. Kalin, "A Graph Generation Software Package," *Proceedings of the twenty-second ACM SIGCSE Technical Symposium on Computer Science Education*, ACM Press, New York, 1991, 151-154.

[JV87] R. Jonker, A. Volgenant, "A Shortest Augmenting Path Algorithm for Dense and Sparse Linear Assignment Problems," *Computing*, Vol. 38, 1987, 325-340.

[K74] A.V. Karzanov, "Determining the Maximal Flow in a Network by the Method of Preflows," *Soviet Mathematics Dolady*, Vol. 15, 1974, 434-437.

[K56] J.B. Kruskal, Jr., "On the Shortest Spanning Subtree of a Graph and the Traveling Salesman Problem," *Proceedings of the American Mathematical Society*, Vol. 7, 1956, 48-50.

[KH80] J.L. Kennington, R.V. Helgason, *Algorithms for Network Programming*, John Wiley & Sons, Inc., 1980.

[KM73] J.L. Kuester, J.H. Mize, *Optimization Techniques with Fortran*, McGraw-Hill Inc., 1973.

[K55] H.W. Kuhn, "The Hungarian Method for the Assignment Problem," *Naval Research Logistics Quarterly*, Vol. 2, 1955, 83-97.

[LD60] A.H. Land, A.G. Doig, "An Automatic Method of Solving Discrete Programming Problems," *Econometrica*, Vol. 28, 1960, 497-520.

[L54] C.E. Lemke, "The Dual Method of Solving the Linear Programming Problem," *Naval Research Logistics Quarterly*, Vol. 1, 1954, 36-47.

[L21] E. Lucas, *Récréations Mathématiques* IV, Paris, 1921.

[M77] S. Martello, "An Algorithm for Finding a Minimal Equivalent Graph of a Strongly-connected Digraph," *Joint National ORSA/TIMS Meeting*, Atlanta, Georgia, November 1977.

[MT85] S. Martello, P. Toth, "Algorithm 632 – A Program for the 0-1 Multiple Knapsack Problem," *ACM Transactions on Mathematical Software*, Vol. 11, No. 2, June 1985, 135-140.

[M57] E.F. Moore, "The Shortest Path through a Maze," Proceedings of an international symposium on the theory of switching, Part II, April 2-5, 1957, *The Annals of the Computation Laboratory of Harvard University*, Vol. 30, Harvard University Press, Cambridge, Massachusetts, 1959, 285-292.

[N66] T.A.J. Nicholson, "Finding the Shortest Route between Two Points in a Network," *The Computer Journal*, Vol. 9, November 1966, 275-280.

[NW75] A. Nijenhuis, H.S. Wilf, *Combinatorial Algorithms*, Academic Press, 1975.

[NW78] A. Nijenhuis, H.S. Wilf, *Combinatorial Algorithms for Computers and Calculators*, 2nd edition, Academic Press, 1978.

[PC79] U. Pape, D. Conradt, "Maximales Matching in Graphen," in *Ausgewählte Operations Research Software in FORTRAN*, edited by H. Späth, R. Oldenburg Verlag, Münich, 1979, 103-114.

[P69] K. Paton, "An Algorithm for Finding a Fundamental Set of Cycles in a Graph," *Communications of the ACM*, Vol. 12, 1969, 514-518.

[P71] K. Paton, "An Algorithm for the Blocks and Cutnodes of a Graph," *Communications of the ACM*, Vol. 14, 1971, 468-475.

[PC71] J.F. Pierce, W.B. Crowston, "Tree-search Algorithms for Quadratic Assignment Problems," *Naval Research Logistics Quarterly*, Vol. 18, 1971, 1-36.

[P57] R.C. Prim, "Shortest Connection Networks and Some Generalizations," *The Bell System Technical Journal*, Vol. 36, 1957, 1389-1401.

[R68] R.C. Read, "An Introduction to Chromatic Polynomials," *Journal of Combinatorial Theory*, Vol. 4, 1968, 52-71.

[S94] M.W.P. Savelsbergh, "Preprocessing and Probing Techniques for Mixed Integer Programming Problems," *ORSA Journal on Computing*, Vol. 6, 1994, 445-454.

[S76] D.R. Shier, "Iterative Methods for Determining the k Shortest Paths in a Network," *Networks*, Vol. 6, 1976, 205-229.

[SDK83] M. Sysło, N. Deo, J.S. Kowalik, *Discrete Optimization Algorithms with Pascal Programs*, Prentice-Hall, Inc., 1983.

[T72] R. Tarjan, "Depth-first Search and Linear Graph Algorithms," *SIAM Journal on Computing*, Vol. 1, 1972, 146-160.

[TS92] R. Thulasiraman, M.N.S. Swamy, *Graphs: Theory and Algorithms*, John Wiley & Sons, Inc., 1992.

[Y71] J.Y. Yen, "Finding the k Shortest Loopless Paths in a Network," *Management Science*, Vol. 17, 1971, 712-716.

[Y72] J.Y. Yen, "Finding the Lengths of all Shortest Paths in n-Node Nonnegative-distance Complete Networks Using $n^3/2$ Additions and n^3 Comparisons," *Journal of the ACM*, Vol. 19, July 1972, 423-424.

[W60] R.J. Walker, "An Enumerative Technique for a Class of Combinatorial Problems," *Proceedings of Symposia in Applied Mathematics*, American Mathematical Society, Providence, Rhode Island, Vol. 10, 1960, 91-94.

[W59] P. Wolfe, "The Simplex Method for Quadratic Programming," *Econometrica*, Vol. 27, Number 3, July 1959, 382-398.

Appendix B: Graph-Theoretic Terms

A *graph* is a set of *nodes* (also called *vertices*) joined by links called *edges*. If e=(p,q) is an edge, then node p and node q are the *end nodes*, and the two nodes *adjacent* to each other. Node p and edge e are *incident* to each other. If p and q are the same node, then edge e is called a *loop*. When two edges are connecting the same end nodes then they are called *parallel edges*. A *simple graph* is a graph that has no loops or parallel edges. Two edges are *adjacent* to each other if they are incident with a common node. A *weighted graph* is a graph in which a number has been assigned to each edge. The *degree* of a node is the number of edges incident to the node. A *complete graph* has an edge between every pair of distinct nodes. If the end nodes of an edge is an ordered pair then the edge is said to be *directed*. In a *directed graph*, or *digraph*, all edges are directed. In an *undirected graph*, edges are not directed. In a *mixed graph* the edges may be directed or undirected. The *complement* F of a graph G is the graph with the set of nodes as G, and two nodes are adjacent in F if and only if they are not adjacent in G.

A *walk* is a sequence of nodes in which two consecutive nodes are the end nodes of an edge in the graph. A *path* is a walk in which all nodes are distinct. A path in a digraph is called a *directed path*. A *cycle* is a walk in which the first and last nodes are the same. An *acyclic graph* is a graph which contains no cycles.

A graph is *connected* if there exists a walk between very pair of nodes, otherwise the graph is *disconnected*. A *strongly connected* graph is a digraph in which every two nodes are joined by a directed path. A *subgraph* of a graph G consists of a subset of nodes and edges of G. A *component* of G is a maximal connected subgraph of G. A subgraph containing all the nodes of G is called a *spanning subgraph* of G. A *tree* is a connected graph with no cycles. A *forest* is a graph without any cycles. A *spanning tree* of a graph G is a tree that is a spanning subgraph of G. A *spanning forest* of a graph G is a forest that is a spanning subgraph of G. A *binary tree* is a tree in which every node has a degree not greater than two. A *cut node* of a graph G is a node whose removal from the graph will increase the number of components of G. A *bridge* is an edge whose removal from the graph will increase the number of components of G. The *node connectivity* of a graph is the minimum number of nodes whose removal from the graph will result in a disconnected graph or a graph with a single node. The *edge connectivity* of a graph is the minimum number of edges whose removal from the graph will result in a disconnected graph or a graph with a single node. A graph is *k-connected* if the node connectivity of the graph is at least k.

Index of Procedures